Is Your Cat Crazy?

Is Your Cat Crazy?

Solutions from the Casebook of a
Cat Therapist

John C. Wright, Ph.D.,
with Judi Wright Lashnits

CASTLE BOOKS

IS YOUR CAT CRAZY

Macmillan Publishing Company   Maxwell Macmillan Canada, Inc.
866 Third Avenue   1200 Eglinton Avenue East, Suite 200
New York, NY 10022   Don Mills, Ontario M3C 3N1

Published by CASTLE BOOKS
A Division of Book Sales, Inc.
114 Northfield Avenue, Edison, New Jersey 08837

ISBN 0-7858-0873-6

Manufactured in the United States of America.
Design by Laura Hammond Hough
Illustrations by Pamela Carroll

To the families who have entrusted me with
the behavioral well-being of their "crazy" cats.
—J.C.W.

To Sassy.
—J.W.L.

Contents

Contents

Acknowledgments

I am grateful to Peter Borchelt, David Hothersall, Katherine Houpt, Molly Robbins, and Victoria Voith for their openness and collegiality during my formative years in applied animal behavior. Their encouragement and suggestions made my transition from a basic researcher to an applied researcher/clinical animal behaviorist a most gratifying experience. Thanks also go to Judi Wright Lashnits and Carol and Richard Wright for providing impetus for writing *Is Your Cat Crazy?* We both wish to thank our editor, Natalie Chapman, and agent, Jim Frenkel, for their guidance and suggestions, and Judi's family, Tom, Erin, and John, for their patience and enthusiasm.

Is Your Cat Crazy?

Introduction

In the past few years, the fifty-five million-plus cats in America have snatched away from dogs the distinction of being the most popular household pet and companion animal in the country. Indeed, the domestic feline—in the form of more than one hundred breeds—is kept and enjoyed by countless millions of people throughout the world.

I'm certain there are lots of perfect pets who bring nothing but joy to their owners and families. And there are many, many more with one or two annoying habits that people just put up with—because Whiskers is so much fun for the kids, or Midnight is so sweet and cuddly, or Tiger is such a good mouser.

But even though their reputation of being "easy keepers"— the ideal low-maintenance pet of the frenetic 1990s—has helped boost the species' popularity, there is a downside to cat ownership. Like dogs, cats are subject to stubborn, maddening behavior problems. And many once-loved pets regretfully—but often with guilty relief—are given away, abandoned, or "put to sleep" because their owners cannot deal with the things their cats and kittens kept doing.

The animal shelters are full of beautiful, healthy cats that very likely could have been lifelong pets had their frustrated owners known what to do when problems arose or how to prevent those problems in the first place. More than four million felines are taken in by shelters every year in the United States. Some are the result of a difficult-to-control population explosion among unowned cats, but many others are there because of unsolved behavior problems that were driving their owners crazy.

Many cats obey several commands, but you can't order them to use the litter box instead of the new carpeting. (Obedience training isn't the answer for this species, any more than it is the total answer for some dogs—as anyone knows who has a canine that heels and sits upon command but still bites Johnny at every opportunity.) These are the same cats that are attacking their owners, spraying on all the sofa cushions, or defecating on the mother-in-law's bed. Or maybe they are acting "jealous" of the baby, scratching the furniture, or picking fights with the dog or other cat in the house.

Until fairly recently, people couldn't help pets with tough behavior problems such as inappropriate elimination or aggression (which make up about 95 percent of my workload) because nobody really knew what to do. Even the so-called experts were just feeling their way along.

The fact is, we are just beginning to get a handle on feline behavior. We still know relatively little about cats, the pets that inhabit at least 20 percent of all households in sixteen countries. Dogs were domesticated several thousand years before cats were, and it's only during the past hundred years or so that cats have been bred in a systematic way.

Many of the dozens of books "explaining" cat behavior are based on casual observations, educated (or uneducated) guesses, and armchair theories, with a number of old wives' tales thrown in;

there have been so few scientific studies of domestic cat behavior that much of what is "known" is actually sheer speculation. As animal behaviorists learn more about typical cat behavior and what makes these fascinating animals tick, we and then the owners will be able to understand and correct behavior problems better.

Only in the past decade have people begun to have access to highly competent and ethical consultants who have devoted many years to learning about the behavior of companion animals. Unfortunately, by the time cat owners turn to professionally certified animal behaviorists—if there's one to be found—they are usually desperate. Our services are often the last resort before they give up on their pets.

They've read everything and tried everything. They've taken suggestions from the veterinarian, the breeder, the local psychic, the next-door neighbor, and the mailman—everyone but an animal behaviorist—mainly because they don't know that we exist. And that's not surprising, considering that there are fewer than fifty certified applied animal behaviorists in North America, pioneers in a still-emerging field.

What exactly do these "cat therapists" do? In my case, I make house calls. I've found it the best way to help the cat owner help the cat.

This is how I work:

When I receive a phone call, usually upon referral from a veterinarian, I generally spend five or ten minutes talking with the caller about the cat and the problem. About one out of five of these people can be helped with some simple advice over the phone. But if the problem continues, or is not one that can be solved with a quick fix (as most of them aren't), I often make an appointment to visit the caller and the cat in their house or apartment.

I make sure the cat has had a recent checkup (by the referring veterinarian, if there is one) before the house visit, to identify any physical problems that may be contributing to the behavior.

I tell the owner to expect a one- to two-hour session, during which I make a behavioral diagnosis and describe and provide a written treatment program. Usually, clients begin seeing results after one or two weeks of the average six-week program and call me as needed for telephone follow-ups until the situation is resolved to their satisfaction.

Once we discuss in the initial phone call the amount of time and effort required by the owner and I am fairly sure about the nature of the problem, we talk about the outlook regarding resolution or management of the misbehavior. If it's clear that the circumstances make the chances very poor, I tell the pet owner, suggest alternatives, and recommend another certified animal behaviorist for a second opinion.

If the chances for improvement are reasonably good or better, we set up an appointment with the understanding that at least some degree of success can be expected. There's no problem behavior that I'll refuse to deal with in cats, but I always talk about options and the likelihood of resolving or decreasing the problem.

But when the first telephone call is truly a last resort, sometimes it comes too late—not for the pet, for you usually *can* teach an old cat new tricks, but for the owner. Some people are so emotionally exhausted, their patience worn so thin, that the prospect of spending six to eight weeks working with their companion animal on a behavior-restructuring program is more than they can handle. And if they can't bring themselves to make that commitment up front, my hands are tied—I can't even try to help them.

How do people react when a vet refers them to a Ph.D. psychologist to help their cats? A lot of cat owners are frustrated and fearful. They don't want to be told that they will have to get rid of their cats, and—because they realize they are all emotionally wrapped up with the cats and the problems—some fear I'll somehow try to take advantage of their situation.

So, if they sound worried or anxious when they call or aloof about the problems, I try to acknowledge that they must care quite a bit about their pets (as I do), or they wouldn't be seeking help. And, although some behavior problems—such as a cat spraying the owner's boyfriend—can be quite amusing to those not involved, I try to take all problem behaviors described to me as seriously as do the owners.

In many cases clients desperately want to make the program work, to enable me to help them. And they want to do their part to help. I'm *not* a "cat therapist to the stars"; most of my clients are middle-class people who come to understand there is no quick fix to most behavior problems. Misbehavior can't be solved by simply throwing money at it or by turning the pet over to a staff of servants.

My cat people want to be responsible for helping the animals they love once they understand that I don't come equipped with a selection of magic spells. Unfortunately, because the cat psychology business is so new, many people have some strange misconceptions about what they're getting into when their veterinarians suggest this type of help.

My phone calls usually go something like this:

"Hi, I have a cat, he's going crazy howling and ripping up the curtains when I leave for work, and I think he needs some—uh—counseling."

It's pretty clear that they expect me to come to their home, "psychoanalyze" the cat, wave a magic wand, and somehow get puss to sit in the window merrily waving his paw as the owner's car heads for the office.

Or: "Dr. Wright, my cat is the best pet in the world with me; she sits on my lap and always comes when I call her. But she attacks our new kitten whenever I'm petting it! Can you make her stop that when I tell her to?"

Some owners can't imagine that anything I might do could make their cats more obedient. But it's not an obedience problem. What has to be done is stop the cat from biting the kitten but not change the other things that make for a great cat the rest of the time. Whether the cat is obedient is, of course, irrelevant.

Or they say, "I've changed my cat's brand of litter twice, I've bought three different boxes, I've put them in four different rooms, I've even shut the cat in the bathroom all day, and she's *still* peeing on top of the piano every day. What would you do that I haven't tried?" (In other words, are you going to come out here, charge me big bucks, and do all the same stuff that has already failed?)

This line of questioning often leads into the guarantee query. Some discouraged owners want to know if results are "guaranteed." (Or their money back, as the saying goes.) The answer is no—it is *never* ethical to guarantee the outcome of this or any other treatment program where an animal's behavior is concerned. As all reputable animal behavior consultants will readily point out, living things will never be as predictable as cleaning products or kitchen gadgets.

I explain that as long as they are willing and able to work with their cats, I'll be available to provide feedback, to make alternative suggestions, and to continue to monitor progress until the problem is resolved, or until the animals are doing as well as they are going to do and they no longer feel the need to check back with me.

When I go to clients' homes, I ask behavior-oriented questions to see what function the problem behaviors serve for the cats (and occasionally, for the owners).

"What did the cat do? What did you do? What did the cat do in response to that? Tell me about the next instance of misbehavior. Then what did the cat do? What made her stop? Oh, she didn't stop—what did she do next?"

And so on. When you concentrate specifically on what the animal *does* and on the people's actions or behaviors that precede and follow the cats' behaviors, then they begin to understand that "animal behaviorist" means just that.

I'm not interested in the animal's psyche; I'm not usually interested in the cat's early kittenhood. We're not going to solve the problem by psychoanalyzing the animal, giving the cat insight into the problem, "breaking" the cat, or making the feline less "schizophrenic" or less "jealous."

Whether the owner thinks the cat is doing "it"—no matter what "it" is—because of spite, or loneliness, or meanness, or boredom, or whatever label is chosen for the misbehavior's motivation, it's what the animal actually *does*, what situations precede the misbehavior and influence its occurrence, and what happens after the episode is over that the owner needs to focus on.

Was the misbehavior rewarding for the cat? Did the feline avoid something even more unpleasant by engaging in the problem behavior? Was the outcome punishment and pain? Was the outcome so upsetting that the cat learned nothing of what the owners intended? By framing the situation this way, I can help pet owners figure out how we are going to go about solving the problems.

What I characterize as results is a reduction of the unwanted behavior from week to week. Occasionally an animal responds immediately, often in the second week, and sometimes not until the third or the fifth week. But as long as we have fewer incidents in the current week than in the previous week, we know the pet is learning something.

If, in two to three weeks, there is almost no change, I have to say to the client (who may beat me to the punch), "This approach is not working; let's try something else." That's where the combination of education and experience comes in—I have to know the professional literature where my colleagues share their research and experiences, and I have to have the resources to try various treatment procedures until the right one for that particular cat is found.

I also have to take into account how comfortable the owners are with adapting to change. If the "best" solutions to their animals' problems won't fit within their particular lifestyle, we try Plan B, or C, and so on. (I recall once being unable to recommend to the referring vet a much-needed tranquilizer for a cat, because the owner was a recovering substance abuser who had just left a supervised group home and was terrified of having any kind of drug in his apartment.)

Very occasionally, I reach a point where I say to clients: We've worked with your cat for seven weeks, he's stopped attacking the kitten except when the little one eats, or stopped defecating in all but one small corner of your house, or stopped biting your son most of the time when he reaches for him. You know and I know that we've done everything possible to help Rambo. Let's discuss whether you can now take what he has to give you, and accept that that's probably the way it's going to be, or whether we should consider another alternative.

In other words, I do my best to enable the client to assess the cat's progress and consider the options that lie ahead. In my fourteen years of consulting, no one has ever had a problem with that. Very occasionally, clients have responded: Well, we can't keep worrying about the kitten, or buying new carpet, or separating our son and the cat forever. What are our options at this point?

I have never advised anyone to euthanatize a cat or dog. It's not my place to do so. But I do talk about the risk of keeping the pet, especially one with aggression problems; the quality of life for the companion animal within the family; and the discouragement that comes with constantly being "in the doghouse." Although I can help provide clients with options and enable them to decide, these are decisions they have to make, knowing that we've given it our best shot. Fortunately, these decisions are the exception.

While I hope to help readers provide the best possible environment for their felines by offering advice and tips for owners, this is not a how-to book for solving tough behavior problems. The course of treatment for each case I talk about in this book was custom-tailored for the particular animal, taking into account many circumstances, including the cat's medical history, age, sex, reproductive status, and breed.

I consider the cat's behavior with its owners and the settings in which the problem occurs—including the presence of other household pets, a new home, a new baby, a restricted or noisy environment, and so on. Each element may contribute to the flavor of the treatment program. And if a treatment option is not likely to be followed by a client because it isn't consistent with the life-style or if, for instance, the client simply refuses to administer drugs to any living thing, then I don't waste time insisting on trying that procedure first.

Occasionally, owners have found that being "creative" instead of following instructions can be dangerous for themselves in regard to aggression problems or can cause the cat's behavoir to regress or the problems to resurface.

Or they may be too close to the problem to see an obvious barrier to success. I remember one client's exasperation with her cat, who would urinate right in front of her anywhere in the condominium without the slightest hesitation or even a guilty look.

"Beauregard is the rudest cat I've ever seen," she wailed on the phone. "Whatever possesses him to do such a thing?" She had no idea what could be causing the problem. I didn't either, until the minute I walked into the apartment and nearly gagged on the pungent odor of peaches and roses and cinnamon and lemon and pine and whatever other fragrance-producing substances were tied up in the cute little packets of potpourri that adorned every table top and sat simmering on the stove.

I wouldn't have been surprised to find plug-in scented night lights in the bathrooms and plastic mushrooms in all the closets. The wrists of my client herself, a sweet elderly antiques collector, shimmered with an overdose of some sort of noxious floral perfume, no doubt recently applied for the benefit of myself and any other expected guests. Poor Beauregard's litter box was full of "pet odor control" potpourri powder. The whole effect was overpowering.

If I hadn't made a house call, I never would have known that the cat probably couldn't smell any of his own waste products, which would have given him a clue where it was appropriate to urinate. Every place smelled the same (beautiful) so it was no wonder he went on the rug in front of his owner with an innocent look on his face. Any place would do. But the owner didn't have a clue that she had gone overboard with the fragrances until I pointed it out, as tactfully as I could. Just phoning me would never have solved her problem.

Another client called me when his cat began urinating on $2,000 worth of newly installed carpeting. He and his wife were understandably frantic.

When I walked into the house I saw brown plastic garbage bags all over the floor on top of the beautiful new white broadloom. "This is interesting," I commented. "Are you trying to keep your cat off the carpet?"

He shot me a look. "Well, you should know!" he replied. "I got it from you on the radio."

"You did?" I asked, racking my brain for when I might have suggested such a ridiculous thing.

"And it doesn't work. The cat just scrunches the bags up and goes on the rug anyway."

"So I see," I murmured.

A bit later in the conversation it became clear that the man had heard me mention using plastic carpet runners during a call-in radio program, but all that stuck in his mind was the word "plastic." I now had to explain to my client that not just any plastic would be helpful in this situation. After he installed some heavy plastic carpet runners with little "feet" the cat couldn't mess with, the carpeted area became protected and unattractive to the cat (but more attractive to people) while we worked on the behavior problem.

There are definite limits to what one can glean from a radio program, a telephone consultation, a videotape, or a book—this or any other—when it comes to a specific cat's behavior problems. If you are the owner of a misbehaving feline, please don't try to solve tricky behavior problems—especially those involving aggression—without expert help. Do consult a professionally certified animal behaviorist, if you have access to one, not only for biting or feline elimination problems and so forth, but also if you are planning a family or if you already have other companion animals or a young child and want to introduce a cat or dog into the household.

What it is possible to take away from this book is a greater awareness of the general principles of cat behavior as a companion animal, along with the knowledge that you *can* go home at night (people whose cats have persistent litter-box problems often seem to dread this) and that you don't have to physically punish or hurt your cat or shut the animal in the bathroom all day or take turns missing work to stay home on guard or give the cat up.

People are extremely pleased when I tell them that there is hope, and that their beloved cat is not "bad," only *doing* something that is not good. So it is the behavior that we need to change, not the pet's personality.

"When she is through with this program, will she be different?" they often ask, wondering if this shrink is going to turn their pets into vegetables, so overcontrolled that the owners won't recognize them.

My usual response, borrowed from my colleague and friend Peter Borchelt, is, "Well, she'll be different in that she won't bite people [or spray all over the house, or fight with her littermate]. If that's part of the animal that you don't mind losing, that's fine. The likelihood is that when she greets you, she'll be playful, confident, and more fun to be around." Pets with problems often don't seem very happy.

I recently got a note from a client making a final report on a pet's progress. It read:

> *Dear Dr. Wright,*
>
> *Thank you for helping me out with Winnie's problem! . . . It's amazing how smart animals are. I believed Winnie was hopeless, but thanks to you . . . she seems to be on the road to being "normal."*
>
> > *Best wishes,*
> > *B.K. and Winnie.*

That's the kind of response that makes everyone's time and effort worthwhile.

The other day I saw an article in a newspaper from one of the world's most sophisticated cities. It contained the mind-boggling statistics that 70,000 animals had ended up at the city's SPCA in 1990, and 5,000 of them were "put down" at the owners' request. Here are some of the reasons pet owners gave for giving their cats and dogs an almost certain death sentence (only 17,000 were adopted):

"I'm allergic."

"My apartment's too small."

"I'm moving."

"I'm going on vacation."

My practice, and this book, are a tribute to those cat owners who don't consider a trip to the Bahamas a good reason to have a companion animal put to death. Before they find help their homes may look like battle zones or smell to high heaven; their cats may be hated by the neighbors; their limbs may be covered with scratches and bites; their patience and resources may be nearly exhausted.

But here are the people who cared enough to give their pets one more chance, and the cats that made it all worthwhile.

John C. Wright, Ph.D.
Spring 1994

1

In Search of the Perfect Cat

The first thing I noticed as I entered Meredith Spencer's tidy garden apartment was the interesting series of small bumping, scratching, skidding, and galloping noises coming from within. Sounds such as these are generally traceable either to exuberant toddlers or to wild-eyed young cats, or both. In this case, the source of the racket was Whiskers, a nearly two-year-old female feline whose kittenish zeal should have been exhausted long ago. Whiskers had been the subject of Meredith's desperate telephone call two days earlier.

"Is my cat crazy, Dr. Wright?" she asked after giving me several alarming examples of her cat's weird behavior.

"Not at all," I assured Meredith. "But let me come out and see you and Whiskers, and we'll put together some changes that should help her."

Now, as I entered Meredith's living room, Exhibit A presented herself. There, racing around like a whirling dervish, was what turned out to be the liveliest cat I have ever encountered. The animal zoomed back and forth across the small living room, leaped and scratched her way up the curtains nearly to the ceiling, jumped down, slipped and skidded across the wood floors and—with a flashy big-league slide—came to an abrupt stop on the braided rug. Then she was off again in a gray-and-white blur, literally climbing the walls, whizzing around like a thing possessed. The cat took no notice of her owner or of the stranger who had come there to gape at her.

Most cat owners know (or find out) that a kitten will occasionally jet around like this in short bursts of excess energy. But if that kind of behavior regularly went on past kittenhood, I seriously doubt that cats would be the most popular companion animal in America. During our ninety-minute consultation that day, my client's cat never so much as sat down—not even once. It was nerve-racking, to say the least.

Meredith gestured helplessly toward her frantic feline. "Meet Whiskers."

"I see what you mean," I said.

Meredith had a number of concerns about her pet; unfortunately, the cat's unusual energy level was only the beginning. Beyond being totally wired, Whiskers had two serious behavior problems her owner wanted very much to resolve: tail biting (her own) and ankle-biting (Meredith's).

The cat frequently pounced on her owner as she walked by, inflicting painful bites and scratches. All this had been going on for two or three months (for reasons that were to remain a mystery), and Meredith had finally had enough.

Although most cats sleep up to 80 percent of the time, Meredith told me that Whiskers dashed around and played all day, rarely lying down at all until she and her husband went to bed. I believed her. But by 5 A.M.—often earlier—Whiskers started bouncing off the walls and had to be shut out of the bedroom so that Meredith and her husband, Larry, and another cat, Monster (a well-behaved, lethargic male), could get some more shut-eye.

It wasn't the first time I had heard complaints of excessive cat activity at night. Of course, if your cat selected 3 A.M. as the perfect time to bat around that earring you lost under the bed, you'd probably characterize the activity as excessive, too. When you can barely open your eyes in the morning, it's hard to understand how a cat can have so much energy in the middle of the night.

The cat's big, beautiful eyes, with their ability to gather lots of light, offer a rationale for nocturnal feline friskiness. The cat doesn't just shut down at night as humans (and most dogs) do. To a cat, there's not a lot of difference between night and day, except that the humans they live with are inactive in the dark. At one-tenth the light level we use to grope our way to the bathroom, our purring pets can inspect the premises after lights-out without any problem. Although they can't see a lot of detail in the dark, they sure aren't going to be bumping into the dining room table.

When felines do rest at night, their sleep takes the form of—what else—catnaps. And while their eyes may be closed (although they can sleep with their eyes open), their senses are apt to be on the alert— at least to those things that interest them. Cats can filter out unimportant sounds while they're sleeping, yet be instantly awake if they hear, say, the patter of little rodent feet. A dog is more likely to sleep right through that and everything else at night. So the smile-provoking idea of a "watch cat" might not be so ludicrous after all.

During our meeting, Monster snoozed on the bed while his female companion flew in and out of every room. Monster, too, had been on the receiving end of little Whiskers's attacks, the owner told me.

But for the most part, Whiskers just leaped around or ran in circles chasing her tail, occasionally biting it, Meredith reported. This event was immediately followed by surprised and indignant howls of pain. The tail chasing went on at all hours, the owner said—at least ten times a day. The only way she had been able to calm down the cat was to sing to her, and Meredith was running out of lullabies— not to mention patience.

I quickly saw that this was not a case of classic tail-biting, in which the cat will systematically chew her tail almost down to the vertebrae. That horrifying behavior is rare. Fortunately, Whiskers wasn't hurting herself deliberately. It was more a case of catching sight of her tail as she whirled around and trying to grab it just because it was there. In her aroused state, she probably didn't even recognize the tail as part of her own body. She would even growl and hiss at it, according to her owner.

"And when she's not chasing her tail, Whiskers is biting or scratching or pouncing on me!" Meredith complained, dodging the fur missile as it hurled by. "She hides in the bathroom at night when I'm getting ready for bed and jumps at my feet when I come around the corner. Or anytime I'm running past her, she attacks."

"And what do you do when this happens?" I asked.

"Well, I shake her off my ankle or run to get away or hide my feet under the bed . . . things like that."

"OK." I made a note. All of these very natural responses unfortunately were the type that would encourage Whiskers to continue attacking. At least Meredith hadn't retaliated and made the cat fearful, for which I applauded her. It's not always easy to resist "disciplining" a pet that has hurt you.

Of course, Whiskers hadn't meant to hurt Meredith at all, any more than she meant to bite her own tail. The cat's frequent ambushes were part of playfighting behavior, in which the victim unwittingly drew aggressive play from the cat just by moving across the feline's field of vision.

Cats' play is made up of stalking, chasing, and attacking—in which they pounce and bite. They do this with other young cats, and if Monster hadn't been such a couch potato, Whiskers very likely would have played with him more, instead of attacking Meredith at every opportunity.

The aggressive pet may have a threatening, scary look, but if the attacks are generally silent, the bites inhibited—that is, not the hardest the cat could chomp down if desired—and the ambushes tend to take place under predictable circumstances—for instance, every time the owner steps out of the shower—we can pretty safely characterize the type of aggression as "playful." And play, in these cases, is a big part of the solution.

I could see that Whiskers was indeed very playful and would chase just about anything that moved. The sorry state of her tail proved that.

"She used to love to play with me," Meredith confirmed. "She would come when I whistled and chase and fetch a ball. But lately she's been too wound up to do much playing."

I explained that play would be key in Whiskers' return to normalcy, and I had Meredith make a list of toys and foods Whiskers especially liked. These would be used to draw the cat's attention away from her tail or Meredith's ankle or Monster's ears—another favorite target.

I told Meredith to keep the treats on hand and put the toys in various accessible locations around the apartment and in her pockets. When it looked as if Whiskers was becoming too hyper and was gearing up for an attack—or was starting to pay attention to her tail—Meredith was to toss a toy away from, but across and in front of, the cat, more or less as bait.

The reason this simple technique works reliably to redirect the cat's attention from the preferred target is that cats' brains have receptors that work like motion detectors to help them see the slightest movement of an animal when they are out stalking their prey. The way their brains function makes cats not only expert at noticing movement but much more likely to attack if objects are thrown away from them at some sort of an angle, rather than dropped directly in front of them.

I explained to Meredith that this kind of horizontal movement is a strong stimulus and would prompt Whiskers to redirect the attack from her to the toy she threw. "It'll be very difficult for Whiskers to wait for your ankles to come by, instead of doing the catlike thing and pouncing on what you've thrown. But if she does happen to catch you by surprise in a sneak attack, try not to encourage her by flailing around—she'll interpret that as a moving stimulus that needs to be playfully attacked."

"So what should I do?" Meredith asked.

"Just remain still if you can or quietly walk—don't run—away from her."

Finally, since Whiskers's activity level was so out of kilter, I asked Meredith if she would consider my discussing the administration of a tranquilizer with the cat's veterinarian.

She frowned. "I'm not sure how I feel about that," she admitted. "I don't want a pet that's hooked on drugs. Is a tranquilizer really necessary?"

I understood her concern and explained that the drug was not meant to be a crutch, or even a solution in itself, but a temporary tool that would help Whiskers become calm enough to learn her new behaviors. I assured her I'd ask the vet to prescribe it in decreasing doses, and if all went well, it would be finished in less than a month.

Meredith still looked doubtful. "Will Whiskers be lying around like a zombie?" she asked.

"Not at all," I assured her. "No one wants a pet to be drugged, but for certain types of behavior problems, medication can be very helpful to animals, just as it can be to people. I'll ask the vet to call you to answer any questions you might have about medication." I suggest drugs or hormones as an option only with the agreement of the veterinarian and the owner, after a discussion of the risks and benefits involved.

With the basic program outlined and in place, we agreed to weekly follow-up phone calls to discuss Whiskers's progress. Meredith's veterinarian called to explain the benefits and risks associated with the drugs and was able to allay Meredith's fears.

What usually happens when we treat playfighting with playing—and this case was no exception—is that as the weeks go on, the cat develops a favorite plaything in place of the attack object. Meredith soon told me that Whiskers now had two favorite toys: a Ping-Pong ball and a slightly smaller yarn ball.

Their play sessions took place fifteen or twenty times a day for only a minute or two in the first week. The cat started batting toys all over the place, and this activity gradually took the place of both jumping on her owner and chasing her tail.

Over the six- or seven-week period that we worked together, Whiskers calmed considerably—even after the drug was discontinued—and Meredith was eventually able to play normally with her cat without being attacked. She even taught Whiskers to sit quietly beside her in a "no-play zone" on the living room sofa and be stroked and petted, something Whiskers had not sat still for in some time.

Now it was time for a revised play schedule. To keep the cat on an even keel, about three times a day Meredith or her husband would play with Whiskers for a period of five to ten minutes. The cat got rid of her excess energy, Meredith got rid of her scratches and bites, and soon Whiskers became a completely civilized feline—right down to the tip of her rapidly recuperating tail.

This cat case is fairly typical of those I handle daily. A cat often starts out well behaved, and then something happens. It can be a sudden trauma, the onset of sexual maturity with hormonal changes affecting behavior, or a slow, gradual change of preferences that no one notices as a problem—for a while. Then, when the cat is defecating on the bed each night or attacking housemates every other day, the frantic owner calls the veterinarian. And the veterinarian refers him to me. By far the greatest number of cat clients come to me with litter-box problems. Aggression is the next most troublesome area, followed by the other disturbing behaviors I'll talk about in this book.

But it's the owner—not I—who holds the key to the pet's

progress. Meredith's cat Whiskers would not have learned her new behaviors had the owner not been willing to play with her cat twenty times a day for a week and trust that prescribed medication wouldn't change Whiskers for the worse.

None of the cats I encounter are "crazy." But most of them need the commitment from an owner or a family willing and able to work consistently to solve the problem and to make some changes in the cat's environment, as Meredith did with Whiskers. When that happens, the treatment program usually yields a happy ending for cat and household alike.

But let's go back to the beginning—before things start to go wrong. Although sometimes the cat picks us—what cat lover can resist the stray kitten that turns up at the doorstep, purring for a handout and a place to sleep?—most of the time we choose our cats. And we do so with our own personal vision of "the perfect cat" in mind. We may know logically that such a creature doesn't exist, but still, there's that special way we want our cat to look or to act, and that's what we set out to find—whether it's from a backyard breeder or the local humane organization or a pet store.

I am no exception. After beginning as a dog owner and dog behaviorist, I found that by the mid-1980s more and more of my clientele consisted of cats and cat owners. I had lost my faithful canine companion Zuk to old age and had been living alone for several years near the university where I teach psychology. Now, I was anxious to become more intimately familiar with these wonderful cat creatures.

So one day I set out to adopt a kitten from a local shelter. Such was my zeal that I went home with the only healthy feline the shelter had in residence that day; never mind that it was no kitten but a six- or seven-month-old cat, personality and temperament a question mark.

I knew that the crucial early learning period was long gone; I had no idea of the cat's background, only that it was a stray. Nevertheless, I was sure that I would be rewarded by having the perfect companion, merging gracefully into my life-style and residing with me in mutual devotion. Any initial adjustment difficulties would, of course, be recognized and resolved smoothly.

My particular idea of the perfect cat went something like this:

Physically, I admired a large cat, a short-haired one that wouldn't shed, probably a neutered male. The breed and the color weren't important. But naturally, his personality and behavior were.

He'd like to sit in my lap. He would seek and keep track of things outside and lie on windowsills and look out. He'd watch me and my activities and sometimes follow me from room to room. He'd feel comfortable sharing the foot of my bed, or even a pillow, at night.

He would possess the pride that is the essence of a cat and refuse to do tricks or suck up to me for any reason. But, on his own terms, he would frequently choose to jump up and sit with me and allow me the privilege of petting him.

My fantasy cat demanded a certain amount of respect and responsiveness from his owner. If I wasn't doing my job playing with him or feeding him or changing his box, he'd let me know. He'd like to go out on the deck and would venture outside now and then to explore our yard or sit on the step and watch things that interested him. If the other cats in the neighborhood should happen to come around, they would see him and scamper off (which is highly unlikely in real life). He would be big enough that no cat would try to mess with him. He would not be a bully, but if another cat foolishly chose to tangle with him, the animal would soon slink off in defeat.

When people came to visit in our home, he'd feel comfortable coming in and sitting down on a soft pillow, observing them from a distance while they admired his beautiful coat and dignified air. He would be pretty much a one-person cat but would graciously tolerate the presence of other humans. Needless to say, he'd never exhibit any of the misbehavior that has less fortunate cat owners tearing their hair out and calling in animal behavior consultants.

The perfect cat.

But back to reality. My little shelter cat turned out to be a lot different from the animal I had invented in my head. She was a small, female black-and-white mixed breed with frightened green eyes. I had her spayed and named her Domino.

Poor Domino soon revealed she was afraid of just about everything. If I looked at her for more than two seconds—which cats do to each other as a threat, with the submissive one usually looking away—she gave a terrified cry and shot out of the room.

Every now and then she came up and started rubbing all over the furniture and let me stroke her as she walked by, but I could not pick her up. If I did, she struggled to get loose, and I knew I wouldn't see her for an hour. This was not an adjustment problem but a temperament problem. For whatever reason—genetics, early experience, maybe some unknown trauma—this cat would not come near me. To this day I am still waiting for her to jump up on my lap, but as far as she's concerned, I can wait forever.

Domino must have had some kind of frightening experience with paper. She is scared to death of any kind of paper rustling sound. I tried to feed her by hand to get her close, but if there was even a little napkin movement she'd disappear in a flash. She is the timid personality taken to its extreme. Even after months in her new home, Domino never let down her guard.

So, as I helped more and more clients resolve their cat behavior problems, I was coming home each night to a companion animal that hid under the bed when she saw me. It was pretty discouraging.

After about a year of this sort of thing, a fellow psychologist suggested I get a second cat to keep Domino company. This time, we would go together to a breeder and pick out just the right kitten. We would be professional and scientific; we were two Ph.D.s with decades of higher education and teaching between us. We could do this. Domino would meet her new feline friend, loosen up, and model the new cat's behavior, and we'd all live happily ever after.

My fantasy cat wouldn't have been a very good match for Domino, so we picked out a pretty calico kitten that seemed assertive and called her Turk. She turned out to be very exploratory and playful, soliciting play night and day. But *she* wouldn't sit on my lap, either. Even now, several years later, Turk will lie on the back of the couch and tolerate stroking, but she doesn't want to be picked up or handled.

In utter disregard of my wishes, the two cats refused to have much to do with one another. Turk, the little newcomer, quickly established herself as queen of the castle. Domino deferred to her under all circumstances, letting Turk have first choice of resting places and waiting until Turk finished eating before approaching the dish.

At least now both cats sleep on my bed. But Domino specializes in kneading on my neck or in my armpit in the middle of the night with an occasional lick of my nose for variety. Then she proceeds to walk all over me. (Turk, on the other hand, fastidiously avoids stepping on any part of my body when I'm in bed.) I usually end up "sleeping" in a posture of near paralysis, afraid that if I move a centimeter I'll send Domino shooting out of the room in a state of terror. And so it goes.

Although my background helps me deal with specific behavior problems that crop up now and then, and I try to be sensitive to helping my two cats cope with the stresses of life in our family, I am not going to change their individual personalities. Maybe they're not the perfect pets, but I love them just the same.

In spite of my own personal record, people often ask me how to spot the "best" animal in a litter of mewing, fluffy kittens, each one more adorable than the next. Should they take the first one that walks up to them because the kitten seems curious and assertive? The shy one that hangs back and looks in need of a friend? The "Socks" twin for star appeal? The one curled up in the corner, looking in need of a lap to sit in? The one whose markings will perfectly complement the decorator's color scheme in the living room? The rough-and-tumble one that looks able to withstand the enthusiasm of two preschool kids?

Obviously, everyone's idea of the perfect cat is different. And there's no real way of knowing how any particular cat is going to turn out, even if you're sure when you scrutinize the litter. But you can control some of the variables when choosing your new friend.

Do you want a petite feline, or is your heart set on one that's big and hefty? If size is important, you should be aware that males are fifty percent heavier than females. If it's the hair that's crucial, you might want to look to the various breeds characterized by long, short, or nonallergenic coats.

Some people want to own cats primarily for purposes of breeding and showing. They will best be served by familiarizing themselves with the characteristics of several breeds through reading, visiting breeders, and attending cat shows where the champions are on display, then settling on the breed they find the most exciting and pleasing.

As for behavioral characteristics, different breeds are often said to be generally active, friendly, playful, intelligent, and so forth. But be careful here—when you go beyond the physical standards, there are always variations within a breed.

Some behaviorists find that pedigreed cats have a slightly higher incidence of behavior problems and that the well-blended gene pool of the alley cat—like that of the mongrel dog—yields a more behaviorally healthy animal. This cat would likely display "heterosis"—a term sometimes used to describe hybrid vigor, the tendency of mixed-gene organisms to thrive in comparison with relatively inbred animals. In my own practice, I have encountered a variety of behavior problems in all types of felines, from barn cats to purebreds of many different stripes.

On occasion, I meet someone who has chosen a breed as a pet simply for its snob appeal. A Flame Point Himalayan on the settee or an Ocicat in the sunroom may be as important to some cat owners as displaying the currently fashionable luxury sports car in the driveway. But generally, whether their pets carry a fancy title or not, most people decide to keep a cat to be a household companion, to love and to care for and—if they're lucky—to cuddle up with on cold or lonely nights.

What seems clear is that the most significant factor affecting a cat's behavior toward humans is its personality—more than the sex or age of the cat or anything having to do with the person. Of course, this influence can be skewed by stress, punishment, or trauma.

We still don't know very much about predicting cat behavior and personality—many of the pioneering studies in this area have been done only in the past five to seven years. But if you're more concerned with how the cat acts rather than the right markings or the proper pedigree, behaviorists suggest some typical cat personality traits that you can use as guidelines when adopting an animal.

One is *alertness*. A cat can be active and/or curious. If she is active, she zips and moves around a lot. If she is curious, she approaches and explores changes in her environment.

A second trait is *sociability*. With people, a cat can be sociable, fearful, hostile, or tense (very careful in movement).

The third trait is *equability*. Does he get along well with other cats? Does he interact with others easily and calmly or not?

Animal behaviorists have identified two types of friendly personality in cats: One prefers play, the other prefers petting. Still other researchers have developed three categories of behavior toward unfamiliar people: initiative friendly; reserved friendly, depending on whether the cat initiates interaction with the human; and unfriendly or fearful. That last is easy to spot—just look under the couch!

When selecting a cat, try to determine whether she seems to like being around people, then observe her behavior style: is she playful or a lap sitter? And if you expect to have another cat (or already have), consider the adoptee's equability.

Bearing in mind the personality and temperament differences in cats, what if you want to get two of them? Nearly half of all homes with cats have two or more, after all. Should they be alike or opposites? What about gender?

From a behaviorist's point of view, we just haven't studied various cat combinations in households extensively enough to shed much light in this area. So you'll just have to blunder along, as I did with Domino and Turk.

Neutering can obviously change relationships, sometimes in subtle, nonsexual ways. Some humane societies are neutering kittens very early in life—at about ten weeks—but the behavioral effects of such early prepubertal alteration are not yet known.

A couple of females are likely to get along as well as would two males. Females are more likely to solicit rubbing than males do, so if you want cats that rub, get females or a female/male pair. If you want your kitten to play with other cats, rear him with others. Kittens are a lot more playful if they have siblings to roll around and play-fight with.

If you plan to adopt more than one kitten or if you already have a resident cat, you probably want to stay with the same sex: If you start with three females, chances are you'll have more success with a fourth female than if you brought in a male, which would provide more competition in the same housing area.

In people, psychologists have found that similarities attract initially, but what keeps relationships together for the long term is the complementarity of needs. So if you generalize that to cats—this is probably a stretch—you could say that if the cats both love playing, but one loves chasing and the other loves running away, their needs are complementary and the two might be a compatible pair.

And if one enjoys playing a lot and the other enjoys just sitting around, you can probably assume that the two of them are not going to interact a great deal. But for the owner who likes to both sit and play, this relationship might turn out fine.

One word of caution. In Chapter 7, you'll find some of the grim results that can come about when there are too many cats—and that can mean as few as two—so be very careful not to take home two or three impulsively unless you're prepared to cope with the consequences.

Cats' personalities probably vary as much as do their human owners'—from timid and flighty to sweet and affectionate to aggressive and sometimes downright mean. How they get that way depends mostly on their genetic makeup and their maternal and other early experiences in the litter, so that by the time you set eyes on the three-month-old kittens at the home of a breeder or your local humane society or a neighbor's garage, they are well on the way to exhibiting the disposition that will characterize their behavior for the rest of their lives.

I remember one cat I worked with whose unusually assertive personality got him in a lot of hot water. Pumpkin bit his owners when he sat on their laps, although nothing seemed to have happened to make the neutered orange two-year-old aggressive. He was a feline who had to have his own way, a bit of a bully, according to his mild owners, Phil and Evelyn Randolph.

If they dared to cross him, Mr. Randolph told me when we met, Pumpkin would get a wild look and bite their hands, legs, and arms. In fact, at that very moment, I noticed Pumpkin shooting a baleful look in our direction. I wouldn't want to get on his bad side myself, I decided as the cat stalked from the room.

Pumpkin did not care to be picked up, thank you, and if he was asked to move away from the front door or do anything else not to his liking—such as having his ears examined (which, in all fairness, no cat is fond of)—he would retaliate. If he did deign to sit on their laps, the couple's pleasure was often cut short when Pumpkin chomped down on an arm, then calmly marched away to go about his business.

Mr. and Mrs. Randolph agreed that Pumpkin *always* retaliated when they tried to influence his behavior. So it was fairly clear to me that Pumpkin's behavioral problems stemmed from a personality trait of overassertiveness, desiring to be in control as "the boss." I felt I needed to help establish a little more leadership on the part of his victims, Mr. and Mrs. Randolph.

Pumpkin's tendency toward assertiveness could be rechanneled or influenced in various ways. Mrs. Randolph, who admitted to being Pumpkin's favorite target, particularly needed to initiate activities with him rather than just wait for the cat to come up and decide to attack her. She had to make her hands and arms unavailable and be ready to redirect Pumpkin's aggression and energies toward a toy.

Pumpkin's was not a vicious aggression—although he was ornery, I believe he was basically longing for some attention within the context of playful aggression—so we didn't discourage him from attacking things. When he would pounce on something his owners provided (other than their limbs), I asked Mr. and Mrs. Randolph to say—with as much enthusiasm as their mild personalities would allow—"Good Pumpkin!" and pull a catnip mouse or other toy along the floor for him to bite and tumble around with.

We actually had to practice this praising, because the owners were not in the habit of talking to their cat at all. This served the dual purpose of redirecting his aggression and providing some much-needed leadership—and playmates—for Pumpkin.

It was actually easier and more pleasant for my clients to remove their hands and arms from the cat and encourage him to play with a toy than to put up with his behavior or resort to punishing him. Within a few weeks, Pumpkin learned that arms and legs weren't available to bite as often as toys were, and that change cut down on the number of bites his owners received.

Pumpkin gradually assumed his proper role in the household as playful pet, and the owners theirs, as companions and initiators of playful aggression. When Pumpkin was able to release his assertive tendencies on acceptable objects provided by his owners, he became less aggressive to the family members and everyone in his household was a lot happier. Pumpkin's owners did not understand their cat very well. Now they do.

In addition to considering physical and personality characteristics, potential owners might do well to look into the cat's past experiences before choosing one for adoption.

All cats are products of several sources of influence broadly classified as genetics and experience. Knowing or observing the parents might give you a key to understanding what kind of kittens will be produced. But the kind of kittenhood your companion animal has had is often key in predicting what kind of cat will result.

Kittens should not be separated from their mother before seven or eight weeks of age. So if someone offers you a furry bundle at five weeks, don't be in a rush to take it. Those extra few weeks are important for the cat's development. Totally hand-reared kittens often have a harder time breeding or may exhibit other adjustment problems.

Most cats forming a close bond with owners and family members do so because they have learned from an early age to trust and enjoy human companionship. Animal behaviorists are pretty certain that the sensitive formative period in felines occurs between the ages of two and seven weeks, during which handling, or lack of handling, by people can influence the cat to become more or less affectionate as a household companion.

When you were a child, you were probably told not to handle a new litter of kittens. The mother would smell your scent on the newborns and abandon them. Or, worse yet, all the handling would somehow damage the kittens, and they would become weak and sickly and die. That's what I was told, anyway.

And if—God forbid—you did pick one up, it had better be by the scruff of the neck, because "that's the way the mother carries them." Not true. Felines don't seem to like that kind of hold very much. It's better to hold the kitten securely against your body with one hand around the chest and front legs and the other under the hindquarters.

The hands-off injunction probably helped to account for generations of standoffish fraidycats that have given felines a bad name for years.

"Cats are cold and unsociable," non-cat people claim. But kittens handled frequently—in spite of parental admonitions—are more likely to grow up liking and seeking out people for affection. My sister's cat, Sassy, who was born in our cousin Peggy's car and handled incessantly by the family's five little kids, has always been one of the most affectionate and friendly felines I've known.

We don't know how old a cat has to be before her personality is stable; we do know that with early handling the most timid kittens will show a more marked change than already extroverted ones. So go ahead, pick up that kitten. If you wait until later, it's more difficult to influence a cat to be more affectionate. If the mother is near-by, she may try to decrease the amount of handling. And she may try to hide the kittens, perhaps to protect them from getting used to another species.

If kittens have been handled during the early important weeks but are then sent to languish in a pet store for several more before being purchased, precious socialization time is lost. Cats must continue to be handled after the sensitive period in order to make the most of their potential as friendly pets.

Most kittens and cats are likely to adjust successfully to living with humans—whether they were born in the closet or adopted and brought to a new home—if you plan ahead before setting out to visit the breeder or the local humane society with your cat carrier. Even when the mother and the kitten are kept together, problems can develop if the environment is chaotic. Many cats have a tough time adjusting to an unpredictable life-style, whatever their age.

The mother-daughter duo of Cricket and Dice, a couple of Himalayans I saw several years ago, would probably be called a "dysfunctional" cat family today.

I was called in by their owner, Eric Cranston, for a litter-box problem involving the kitten, Dice, who had just defecated on the new couch. This snowy kitty with black splotches on her tail got off to a very shaky start from day one, thanks to an anxious mother facing an environmental overload.

Eric, a sociable young college student, lived alone but had a constant stream of friends coming and going—one of whom left him with a pregnant Cricket while she went off to study abroad for a year. Dice was born in the bedroom closet, where she lived for a time with her mother and several littermates. When they outgrew that space, the kittens tumbled out into the living room, accompanied by the mother cat. That's where the trouble began.

After the other kittens went one by one to new homes, Cricket and Dice were introduced to a disruptive visitor. They had to share the apartment with a difficult guest cat, a sly female tortoiseshell that stayed (along with her owner) for several long weeks. The guest cat, Eric said, bit and aggravated first-time mother Cricket, who did her best to raise her kitten while dodging the intruding cat.

When we sat down together, my client told me that the mother cat had been very nervous and had nursed Dice for about three months, well beyond the average time. By the time she was weaned, Dice was having problems using the litter box consistently.

Finally the guest cat left, but before the two felines could breathe a sigh of relief, Eric took the whole show on the road. Lacking a catsitter, he packed up his cats and drove from Georgia to Florida for a week and then packed up again and took them back to Georgia. A month later, he took them to Florida for another week and then back again. When in Florida, the cats had to stay in a hectic business office, full of turbulent encounters with a variety of cat lovers and haters.

By the time they came home for good, the cats were completely stressed out. In the car, during the last leg of the trip, Dice had attempted to use the litter box, only to have her mother jump in on top of her. This frightened young Dice so much that she urinated all over herself and her mother.

After this unfortunate episode, the litter box at home became even less attractive to the kitten, but didn't stop mother Cricket from trying to "supervise" whenever Dice ventured near one of the four boxes Eric had placed around the apartment.

To make matters worse, Eric had taken to punishing Dice when he discovered an accident. He would spank her and put her into the litter box, thus making it an even more terrifying location than ever—and very likely leading to more accidents.

After Eric finished bringing me up-to-date on the kitten's unfortunate history, I asked him where he kept the four litter boxes.

"Well, let's see. Right now there's one in the kitchen and one in the bathroom—no, wait, I moved that one to the back hall, and I think there's one under the dining room table—unless I took that one and put it in the closet. Let me go check. . . ." He zipped around the corner. "Yeah, I did move that one to the closet," he said, trotting back to the living room. "And one in the laundry area."

I set down my pen. "So you move all the boxes?"

"I guess so."

"How often do you move them?"

"Not that often," he said. "I guess I reshuffle them every couple of days. Whenever Dice picks a new place to go, I put a box there."

"I see," I said, picking up my pen. If their owner couldn't keep track of where the boxes were on any given day, I wondered how he could blame the cat for making mistakes. Cats aren't wild about changes in their living areas, especially in the objects that have biological significance—like litter boxes and food bowls.

"I like the cats, but they're really getting out of hand," Eric complained. "The vet said you could probably get Dice to shape up pretty fast."

"I hope so, but you're the one who'll have to work with Dice day by day until her problem is under control."

"Right. . . . My folks are coming to visit in a few weeks," he said nervously. "They're going to send me back to the dorm, man, if they find cat s— all over the place. They don't even like *good* cats."

"Your cats are good; they've just had some hard things to cope with," I assured him. "Let's get to work."

Eric had tried to solve his pets' problems, but he didn't do anything long enough or systematically enough to make any difference, and he didn't have a clue as to what to do next. I explained to him that the cats needed some stability in their lives if they were ever to become calm and well behaved. The punishment and the litter-box roulette would have to stop immediately.

"We'll be making some changes over the next couple of weeks, and after that, Cricket and Dice will have a predictable environment to live in, and you won't feel the need to punish them or follow them around with cat litter," I told him.

"Awesome."

"OK. Now, does Dice ever go when you put her in the box, or does she just jump out?"

Eric thought for a moment. "She goes sometimes."

"And what do you do if that happens?"

"I sometimes say 'cool' or somethin', and then she jumps on my lap and purrs real loud."

"Great! Sounds as if she responds well to praise. Let's keep rewarding her for going in the box. If you'd like, you can bring her to the litter in a calm and quiet way—pet her as you carry her over, so she doesn't think you're going to spank her. If you catch her already using a box, be sure to praise her once she's finished. Make sure you don't startle her while she's in the box."

I told him not to put Dice in the box, which she might be connecting with punishment, but just to set her several feet in front of it and let her step in. I was pleased that the Himalayan was behaving true to her affectionate, people-liking breed. Not all cats would be so agreeable after being spanked for some unfathomable offense. Uptight cats are also likely to reflexively tighten their sphincter muscles, making it impossible to eliminate when unceremoniously plopped into the litter box by well-meaning owners.

I wanted Eric to locate all the boxes well away from the cats' food and water dishes and extracted from him a promise that he would have all the rugs cleaned (the cat never went on hardwood floors). I also recommended that he buy some cat toys and place them in the previously soiled areas. There were plenty of spots to choose from; the cat had gone in every room. The toys were designed to remind Dice to play in those areas, rather than urinate there. To cats, those two different motivations are not compatible, and I hoped to replace Dice's elimination motivation with one for play.

Next, since Eric was very concerned about his couch—his one new piece of furniture—we decided to establish it as a petting zone. This was the place for interaction with Dice, a spot she could use for playing and petting instead of for elimination.

But the heart of the program was to get rid of two of the litter boxes and set up a "zone defense" with the remaining two. I instructed Eric to separate the mother cat from the kitten, especially at night, for the first week or two—until both of them became less anxious and the little one established new litter-box patterns without Cricket's interference.

The first box would go in the bedroom with Dice. The second would be kept in the dining room on a plastic carpet runner. For the first week, the kitten would be kept only in the bathroom and bedroom, which could be closed off by an existing door.

The second week, we would open up access to the hall, then the living room, and gradually the rest of the apartment. Eventually the cats would have the run of the house again. Eric's pets would know where everything was and would remain. They could begin to stabilize their behavior as a new environment, free from chaos, was established. I also suggested to Eric that he start seeking a reliable cat-sitter for those occasions when he would be going out of town, instead of taking the cats with him.

To Eric's credit, he followed the program to the letter. Most important, Eric gained a better understanding of the needs of "college cats"—and though he became no less sociable, he managed to protect his pets from the stresses of living in a popular hangout. Guest cats were no longer welcome at his apartment, and he kept the cats in a stress-free zone away from strangers whenever friends came over.

By the end of his parents' visit five weeks later, Eric told me he had overheard his mother telling Dice that she was a "good kitty," and there was no chance at all of his folks making him move back to the dorm. I shudder to think what their reaction might have been six weeks earlier.

E ric was lucky, and little Dice was fortunate to have the kind of intervention that protected her from long-lasting behavioral problems. The changes we made also helped the mother cat get back on track.

Of course, many of these troubles could have been avoided if Eric had been aware that felines are most likely to adapt to new living arrangements and homes with a few simple guidelines. Here are the steps I recommend when owners are ready to bring a cat into the household.

1. Select a room the kitten can use as home base for the first day to several days. The room should have a door or some means of making the space private and protecting the kitten from the chaos of daily living. This is especially important if the household contains other pets or young children.

2. In a corner opposite the door, place the kitten's litter box, preferably containing the sandy, clumping-type litter the majority of cats seem to prefer. If the kitten has already established a preference for a different type of litter, use it. Buy a litter scooper and resign yourself to using it frequently (at least daily). If this seems like too much trouble, plan on recarpeting the house a few months down the road.

3. In a different corner, or as far away from the litter box as possible, place fresh water and food. Eating and eliminating involve competing motivations for a cat, and keeping dishes and litter boxes well separated will help enable the cat to do the right thing in the right place. You might place a plastic carpet runner beneath both the feeding area and litter box for easier cleaning.

4. Select another corner or area away from the litter box for the cat to sleep; you probably will have to change this location to a self-selected sleeping area later (few kittens prefer to sleep where *you* want them to).

5. Go get the kitten!

I urge potential cat owners to consider adopting from a good animal-control facility or humane society rather than buying from a breeder or pet store, especially if breed and age are not important considerations. According to the American Humane Association's 1990 Animal Shelter Reporting Study, the great majority of cats brought into animal shelters in the United States that year were euthanatized. With the number of cats in America increasing by more than one third during the 1980s, somewhere between 5.7 million and 9.5 million cats were put to death, many for space or housing reasons alone.

Animal shelters charge reasonable adoption fees, frequently make available low-cost spaying and neutering services at the facility or through participating veterinarians, and offer a way to help cut down on the homeless cat population. You may even find entire litters there that have been previously quarantined, and if healthy, placed for adoption.

When a cat is taken in by an animal shelter, he usually will be designated as either adoptable or slated for euthanasia, depending on his health, temperament, age, and other factors. His entry status as a stray or an owner-donated animal will be noted. There often will be a 24-hour period to permit the cat to become accustomed to his surroundings before being shown for adoption. Even so, you should keep in mind that the cat or kitten may still—understandably—be frightened and wary when you examine him.

Try to take several adoptable cats or kittens (no younger than seven weeks old) to a quiet room to observe their play and to see how they react to you. While this may be hard to come by in overcrowded, underfunded shelters, I've been encouraged in recent years to see some beginning to provide nice play areas for their wards.

If you are looking at older cats, don't assume that the one that hides in the back of the cage will necessarily be fine when you get home. No test at the animal shelter can guarantee a good temperament or problem-free behavior, but the cat or kitten that seems to be doing a good job of coping with the stresses of life in the facility will probably best be able to cope with yet another change—going to live with you.

Many felines—such as the ones you'll meet in the next chapters—are disastrously inflexible or easily stressed out. Other kittens appear somewhat inhibited in their behavior at the shelter, yet are real terrors in a quiet home. Still others I've met have appeared relatively active when caged alone at the shelter yet hide under the couch in family homes consisting of several equally active young children. So the particulars of the home situation have to be considered as the search goes on.

The cats may be vocalizing a lot as you examine them. There is no one reason why kittens miaow a lot. But the kitten you're looking at will let you know if she is scared or uncomfortable. If you go toward her and she runs, squat down—don't loom over her. If she extends her claws and lays her ears back when you hold her, the kitten is not happy and should be set down.

One fairly reliable way to get a cat to feel comfortable and approach you is to squat down and stick your finger out as he walks nearby. Allow him to approach and touch his nose to your finger. He will sniff and rub his head on your hand and go on to rub some more and sit by you. Touching noses (your finger is probably perceived as a surrogate nose) is a form of greeting among cats.

Rubbing is a behavior you can also use as a clue to help you tell if a cat likes other cats. Mutual rubbing as a greeting is an outward sign of the social bond among felines. Foreheads, cheeks, and flanks are the primary parts to be rubbed.

When two cats approach each other, one cat raises her tail in the air. If the second cat is friendly and raises the tail in the air as well, you can bet that they are about to simultaneously rub. If the first cat's tail is raised and the other one's isn't (but there is no hostility), the first cat will rub and will keep rubbing until the second cat responds in kind. They do the same thing with people.

You may encounter a kitten that goes around the room rubbing on everything in sight, including the slacks you just got dry-cleaned. "This one," you figure, "just has to be the friendliest."

But hold it—there's another possibility. If a cat is a little afraid of you, he may redirect his rubbing onto something else—such as a chair leg or other vertical object. If he comes up and rubs you, and you stroke him in spite of the dry-cleaning bill, he will stop rubbing sooner. Cats that come in from outside will typically rub their owners. Cats that are around their owners constantly and are very comfortable and not fearful will do less rubbing. So if you think of that very "friendly" cat as a little bit scared but with her heart in the right place, you'll understand her more accurately.

After making the mandatory arrangements for any appropriate veterinary examination, inoculations, and neutering, it's time to bring your kitten or cat home. Ask the shelter or previous owner for a small sample of used litter from the kitten's box to take with you.

You've already gotten the room ready; just don't call the kids in to converge on the newcomer while you go get a cup of coffee. Your cat's first impressions of the new home are important.

Place the kitten in the room with you, with the door closed. Allow the kitten to discover where everything is. *Do not* take her to the various locations. *Do not* place the kitten in the litter box. Just put the sample you brought in the box, and she will get the idea.

Even if you're an assertive, take-charge person, try to restrain yourself. Be passive. Allow the kitten to discover the room's appealing features—windowsills to sit on and look out from, a comfortable couch or bed, tables to jump up on, and anything else in the room that might be useful as part of an escape route, should she need one.

You may wish to leave the kitten alone when she is ready to take a nap or relaxed and confident about the new surroundings. If Kitty chooses to nap in your lap, consider yourself lucky. Of course, you can always try to place the kitten in the sleeping area you've selected and then leave the room.

Your new companion animal will let you know when the time is right to take on the rest of the house and the household. It's important that you allow the kitten to do this at her own speed. I've seen kittens spend as little as a half hour and as long as three days in "their" room. A door opened just enough for the cat to enter or exit is a good starting place for the newcomer to make the transition.

Keep other pets and family members from the cat's room. Introductions to these individuals should be well controlled. Again, allow the kitten to approach and withdraw (back to her room, if necessary) from people or other pets that remain passive. You'll need to control the dog with a leash at first. Kitty will eventually establish the house rules with Fido, but young felines will need help in the beginning. You'll find some cat-dog introduction techniques in Chapter 13.

Once your cat has become familiar with her new home and household members, you'll become aware of her preferences for playing, eating, sleeping, and other activities. At this point (anywhere from a few hours to several days), you may change the location of the food bowl and the litter box to areas preferred by both you and the cat. Let's hope these locations are one and the same, for you definitely don't want to start out with a power struggle involving the crucial litter box: the problem area most likely to send owners to a cat therapist—or their cats back to the shelter.

2

Secrets of Litter-Box Success

"We just had the carpet cleaned, and she's using the whole apartment as a litter box."

"The cat is wetting on all the chairs and sofas. I'm a nervous wreck."

"Kitty sits in the box for fifteen minutes, then goes somewhere else."

"My cat has been urinating outside the box once a day for two years."

"Whenever any visitor goes away after paying attention to him, he poops on the rug."

"The cat got excited about our wedding presents and started going all over the place."

"When our cat is mad at us, he goes on the bed, on my shirt, and in the houseplants."

And so on, and so on, and so on.

The litter box is the biggest disaster area the majority of cat owners encounter. When a cat rejects the litter box in favor of the carpet, the bedspread, or the bathroom sink, it can wreak havoc in the most well-managed households. I've had many desperate clients tell me they literally didn't want to go home at night to face another certain "accident" by their pet.

The problem of cats' failure to use their litter boxes fills animal behaviorists' files to overflowing. Although it is often resolvable with patience and know-how, this distressing habit probably sentences more felines to premature death each year than all other misbehavior combined. And it rarely fails to drive owners up the wall, especially when "rubbing his nose in it," shutting the cat in the bathroom, or using a squirt gun have failed to produce the desired results (which is almost invariably the case).

Litter-box problems have a variety of causes and solutions, and they never just go away without some form of action. Left alone, they can become life-threatening to a cat. Not that it will die from pooping on the sofa, but it is very likely to end up being given to a shelter, euthanatized, or simply let out the back door to fend for itself when the owner's patience and attempts at home remedies run out.

Starting out the right way with the litter box will be your best defense against problems developing later. So what's the drill with the litter box?

Unless certain criteria are met—from the feline point of view— a cat will take its business elsewhere. It's as simple as that. The key here is "the feline's point of view." What cats need for the optimum discharge of bodily functions are three things:

1. Cleanliness
2. Privacy
3. Escape potential

Cats are creatures of habit, and most have already established bathroom routines by the time you meet.

Up until ten days of age, the mother cat has to lick the anogenital area of the kitten to get it to eliminate. It is a reflexive action. The muscles are not well developed, and the mother's lick stimulates the kitten to relieve itself. This is not a distasteful process for the mother, and it actually helps bond the mother to the kitten, because she then knows what her kittens smell and taste like. The kitten also benefits from this process, and eventually its muscles mature so it can go by itself.

When you obtain a new cat, it has already been eliminating somewhere for about six to eight weeks if it's a kitten and much longer if it's full grown. In many cases, that will have been in a standard litter box filled with some kind of commercial litter. But it could also have been in the soil behind the barn, on bare cement inside a cage, on newspapers, and elsewhere.

Some pet stores use cages with wire floors that allow waste to fall into a collection area below. Many indoor-outdoor cats prefer doing their toileting outdoors but will still need an indoor litter box as a backup, especially if they are kept inside at night. It is important to find out exactly what the cat's experience has been in this regard before you bring it home. Some kittens may make a change readily, but others may need a gradual introduction to a "normal" litter box.

Smudge was one cat I met who didn't want anything to do with pans containing granular, malleable stuff for her toileting. That ruled out any kind of litter in a box. What she really craved for that purpose was something hard and smooth. Something like, say, the bathtub.

That's where matters stood when I was summoned.

"We've tried every kind of litter in the supermarket, Dr. Wright," her owner, Sam Peltz, explained patiently as we met at his home. "We just can't seem to find one she likes as well as the bathtub." He shook his head, discouraged.

"Will she go into the litter box?" I asked.

"Yes, she'll step into it," he said.

"But she usually steps right back out," added his wife, Millie, who had just come into the living room with Smudge in her arms. The offending cat was a compact but muscular striped shorthair with little tufts of fur at the tips of her ears and big gold eyes. "Then she shakes her little paw like this [clasping Smudge's paw and scratching the air with it] and runs upstairs to the bathtub. That's when I put her outside."

"Cats usually shake their paws like that when they find something distasteful to their touch," I said. "She really doesn't seem to like litter."

"She never has," Millie added.

"Has Smudge shown any preferences other than the bathtub?"

"Oh yes," Sam answered. "Often she scratches on the outside of the litter box, or on the hood, and then goes right next to the box."

"She's gone on the wood floors in the hallway," his wife added, "and we have to be careful about spreading newspapers anywhere on a floor or table. She seems to like to go on them. Remember what she did on the papers under the Christmas tree?" Millie gave her husband a gentle nudge. He nodded, rolling his eyes. Smudge mewed, jumped down from Millie's arms, and trotted out.

"You say the cat has never liked a litter box," I injected. "Where did you get her?"

"We bought her from a fellow that raises cats," Sam answered.

"Did he use litter boxes, or did he just clean out their cages?"

Sam and Millie exchanged looks. After a minute they replied that neither of them had noticed.

"But I'm pretty sure the kittens didn't go in the bathtub," Sam said.

"We assumed all cats would use a litter box if it was kept clean," admitted Millie.

"Well, most cats will, but it's hard for certain cats to adjust to a new way of doing things if they haven't had a litter box as a kitten,"

I explained. "A lot of pet shops and some breeders just put down papers or use a stainless steel bottom of the cage. It sounds as if Smudge is trying awfully hard to duplicate her early experience of going on some smooth, hard surface."

"What can we do?"

"Well, we know what Smudge doesn't like," I replied. "Let's start with that. If you're willing to put a nice fluffy bath towel or a couple of inches of water in the bathtub when you can't keep the door closed, I bet she'll lose interest in that preferred location."

The couple readily agreed to that measure.

"Obviously, you're going to have to be careful where you leave any newspapers or paper bags," I mentioned as we went to take a look at the litter box.

"We've been pretty careful about that since Christmas." Sam grinned ruefully.

I noted that the location of the litter box, in a little-used sewing room, seemed to offer adequate privacy and escape potential—important ingredients I'll elaborate on shortly.

"We thought Smudgie might like a new box with a lid, but it didn't seem to help any," Millie pointed out. "She's pretty suspicious of the whole affair."

"Let's ditch the cover," I suggested. "They sometimes appeal more to owners than to cats."

Millie took off the lid and stored it in the pantry. "I still have the old box down in the basement."

"Great," I said. "Why don't you bring it up, and Smudge will recognize her old box. This one is still somewhat novel, but we'll use it in an alternate location so your cat will have a choice, and she'll get used to both boxes." Millie agreed.

"Now, let's sit down and work out a way to get Smudge to seek out these litter boxes," I said.

The plan I came up with was designed to make the immediate area surrounding the box unattractive for elimination purposes. To accomplish this, the couple put a carpet remnant under the boxes. This discouraged the cat from beginning a scratching ritual on the outside of the box, followed by urinating or defecating next to it—this cat simply wouldn't go on a rug. (Other problem cats will go *only* on rugs!)

In dealing with the litter box itself, the owners started with a nearly empty pan. A few newspaper strips and a couple of tablespoons of litter were meant to initially attract the cat to use the hard surface, instead of shaking her paw and leaving the coarse, claylike litter in disgust. We figured that the cat would find a sandy, clumping litter the least disagreeable.

Once this good habit was established, the owners would very gradually add more litter and more newspaper each day. Eventually the ratio of litter to newspaper would increase until the newspaper was left out entirely.

Sam and Millie were conscientious partners in this exercise, and everything went pretty much as expected. On the first follow-up phone call a week into the program, Sam noted that with the litter box lid removed, Smudge had tried scratching the wall a couple times. But, perhaps due to the piece of carpeting beneath the box, she had chosen to get into the box to urinate instead.

Within a few weeks it was obvious that Smudge had joined the ranks of the millions of litter-box-trained housecats—and her owners had gratefully reclaimed the master bathtub as their own.

Cats such as Smudge, who had to be trained to use a standard litter box, do not account for a lot of calls to animal behaviorists. Much more frequently, we hear from people whose cats are able to discriminate between an appropriate place and an inappropriate place to eliminate, but for some reason are rejecting the litter box in favor of a better (from their point of view) location or group of locations.

One of the biggest mistakes new owners make—most often

those who have previously been through the horrors of housebreaking a puppy or have never owned a cat—is using physical means to teach the kitten to use the litter box. Although I try never to blame a client for a cat's behavior problem, I cannot overemphasize the need for a hands-off approach to litter boxes for new kittens. Physical contact in the form of punishing, or even placing a cat inside a litter box, can backfire.

Some owners apparently don't believe this. It seems too easy. So when the cat appears ready to go, they will get quite tense, grab the cat, and yelp, *"Over here, over here, over here!"* And then they'll plop the kitten into the box and loom there, urging, *"Go in the box, go in the box!"* Naturally the cat freaks out, jumps out of the box, runs away, goes elsewhere, and never wants to venture near the box again.

Next, the owner finds the spot on the floor where the cat has urinated and, in order to punish and teach the cat that this was the wrong thing to do, rubs the animal's nose in the puddle. Now the cat is terrorized about the urination process, doesn't want to be around the owner, and is very likely to sneak off and do it somewhere inappropriate. The cat may quickly learn to equate seeing the owner and smelling the urine-stained carpet with being punished, because that's what happened the last time. So inadvertently, people who mean well can create big litter-box problems for themselves and their cats.

Generally speaking—and this is one of the joys of cat ownership—you do not need to show a cat how to use a litter box. This is something that cats seem very well prepared to do early in life, even if they haven't seen their mothers do it. And most properly reared kittens have had ample opportunity to imitate the mother, who is already using a box you (or someone) have provided. All you need to do is set the box down in the appropriate place (you can add a small sample of feces from the kitten's previous box) and let the cat discover it, preferably after eating, a lot of exercise, or a long nap.

Next, there is the litter box itself. What kind is best? That depends. If you can find out what sort of box the kitten is used to, getting the same type is the safest bet. The biggest difference among litter boxes is whether or not they are covered with a lid of some sort.

Some cats couldn't care less one way or the other, some prefer a lid, and some won't go near the box unless you take that cover off. If your cat seems comfortable with the first way you try it, leave it.

Now, what should you put into the litter box? Up until a few years ago, the answer to that question was easy—some kind of clay litter. Although some people have always used homemade litter such as soil, torn-up newspapers, and the like, the overwhelming majority went to the grocery store for clay-type litter. This material necessitated frequent changing of the box and was not very effective at masking odors.

In the first few years of my cat treatment programs, I often encountered cats who just didn't seem to like the litter they were using; it may have been the dustiness, the consistency, or who knows what. Owners were trying a different brand week after week, and I believe some of the perfumes and chlorophyll used to make the product smell good to people were driving the cats away from their boxes. (As I mentioned in the introduction to this book, too much odor masking via perfumes, potpourri, and stick-on or plug-in deodorizers can make some cats so confused that they just go anywhere, since everyplace—including the litter box—smells the same.)

Today, we have a good alternative that is rapidly growing in popularity because cats really seem to like to use it. The new sandy clumping-type litters have been shown in behavioral studies to be preferred by cats and are what I now recommend in most cases. (I have no ties to any litter product.) The litter clumps together with moisture, and waste products can be removed daily, leaving the remaining clean litter in the box. The need for perfumes to mask odors seems to be greatly reduced, as the odors are also absorbed.

One of the easiest litter-box problems I was called upon to solve involved a nice gray tiger-striped cat named Lacy. Lacy's owner was too creative to resort to the old shopworn remedies of "rubbing her nose in it" and so forth. John Miller had booby-trapped the apartment to wipe out all opportunities for his cat to leave any surprises (or so he hoped).

His cat was sweet, friendly, and well behaved, except that she had been urinating around the house for the past ten weeks, mostly along the baseboards and on the carpets. When Mr. Miller heard Lacy scratching on the floor—her typical signal that she was about to do her thing—he went on red alert and checked out all the likely areas.

He had found urine under the rolltop desk, so this was one targeted spot. He blocked it with a large chair, which no one was supposed to move. Under the cat's favorite chair in the dining room, he had put a pan of vinegar to deter her.

"Look what I've got over here." Mr. Miller led me rather proudly to his next secured area. Under a handsome Windsor-style rocker in the living room he had built a pile of newspapers topped by some two-sided sticky duct tape. "Lacy doesn't go near *that* chair much anymore," he chortled. (And neither did anyone else, I'll bet.)

"We're still working on what to do when we hear her scratching under the bed," Mr. Miller admitted. "So far, we've just been chasing her out of the room if we catch her under there. I was hoping you'd come up with some new ideas." He rubbed his hands together and looked at me with eager anticipation, visions of brand-new contraptions no doubt dancing in his head.

"Well, you've certainly tried to stay on top of things," I complimented him. "Now, could you show me Lacy's litter box?"

"What? Oh, that. Back in the pantry." He waved an arm toward the kitchen, obviously unthrilled by my mention of this mundane object.

Mr. Miller trailed me to the pantry, and I took a look at the box. It was uncovered, clean, and filled with a popular brand of standard grocery-store clay litter. "Box seems clean enough," I murmured.

"Of course it's clean! She never uses it!" He laughed heartily at his joke. Mr. Miller had a good sense of perspective; Lacy was one cat that obviously never got punished.

"Sometimes she sticks a paw in there, but then she shakes the litter off." He bent down and affectionately stroked his pet, who was curiously nosing around her box.

"Have you ever tried a different type of litter?" I asked when he stopped chuckling.

"Nah, they're all alike," he said.

"Well, actually, they're not," I said. I explained that there were various textures and told him about a recent study by Dr. Peter Borchelt in which nearly one hundred cats were able to try out fourteen types of litter that differed in texture, granularity, or coarseness. By almost three to one, cats preferred the fine, sandy clumping-type litter over other types. (Sand itself seemed to be rather unpopular—it was probably too heavy and densely packed for the cats' liking.) For Lacy, a change of litter was definitely worth a shot. I suggested to Mr. Miller that he try some of that new litter.

"Just change the litter? That's it?" He seemed crestfallen.

"You could also put some toys down in the spots where Lacy has had accidents before, so she thinks of them as play areas," I suggested.

"I'll get right on it!"

"Let me know how it goes with the litter, and we'll take it from there."

As it turned out, my hunch was right. Lacy, a rather dainty little cat, loved the feel of the soft sand on her paws. After refusing to use the box for ten weeks—for what initial reason, we'll never know—she was lured back by the new litter and never had another accident as long as I was in touch with the owner, whose home soon lost the carnival funhouse look. If only all cat behavior problems were so easy to resolve.

« »

Regardless of the type of litter you buy, the litter box must be kept clean or the cat will reject it. Cats have different levels of tolerance for a dirty box. Some won't set foot in it if there is any evidence that it has been used. Others will cautiously step around a week's worth of accumulation and heroically do their duty. But why take a chance? Keep the box clean.

Now that you've got the box, and the litter, where do you put it? Well, how about in a spot that is really out of the way, where company will never see (or smell) it? Somewhere even the cat won't think to look, right? Wrong! The location of the box is one that has to be chosen from the cat's point of view. Don't worry; it doesn't have to end up in the living room—although I've had several clients who chose just that solution.

Buster, like most cats, needed cleanliness, privacy, and escape potential to solve his box problem. But he had his own definitions of the three ingredients.

As to cleanliness, Buster wasn't particularly fussy; his box was cleaned out a couple of times a week, and that was good enough for him. But when his family went to the beach and hired a neighborhood kid to come in and feed the animals—of which there were plenty—changing the litter was something the catsitter apparently forgot to write down. Toward the end of the vacation, Buster finally rebelled.

After two weeks, the owners returned to a well-fed feline that had started leaving unpleasant surprises around the house as an alternative to using his now-filthy box. The catsitter was angrily crossed off the list; the owners dumped the sand out of their shoes, poured Buster's into his box, and crossed their fingers. But the feline turned up his nose, marched away from the box, and did his business on the carpet. They put a second box near the first one, but he ignored it. That's when my phone started ringing.

When I visited the home of Mabel and Fred Thompson, I was struck by what a zoo it was. Family members and visitors were everywhere, there were dogs and rabbits in and out of the house, and everyone and everything got along marvelously. Through it all, Buster seemed remarkably laid back. Although his demeanor was calm, his dilated eyes indicated to me that he was aroused and perhaps under stress.

The family told me that when Buster failed to use the box they would go get him and plop him in it. So he was probably now associating the box with being punished and avoiding it all the more. (Sometimes he would enter the box, and the family would reward him with a cat treat, but this happened rarely.) I noted that medication might eventually be needed to relax Buster in this overly stimulating environment. But I wouldn't consider that option until I had first tried to help the family make the cat's surroundings more cat friendly.

The litter box he had originally rejected was in a pretty disastrous location—the laundry room. There was only one way in and out, and with all the traffic he encountered daily, Buster would have been much better able to get back to using a box if it afforded more escape potential.

Buster was unusual in one regard—he didn't seem to be fussy about eliminating near his food, which was located near the box in the laundry room. In fact, on one occasion, he defecated right into his food dish. Either the cat didn't care if he did his toileting near his food, or he was too upset to notice. So I knew that separating the box from the food probably wouldn't have the usual effect of solving the "competing motivations" problem.

The second litter box had been placed too close to the first for Buster to be able to discriminate between them. From the cat's point of view, there was only one ideal location for elimination—right in the middle of the very open family room, where he could see things and escape in all directions. He also went along a wall in the living room.

"I don't suppose you would consider putting a box in the family room, Mrs. Thompson," I suggested. Most people don't want cat boxes in their main living areas, regardless of what the family pet has to say.

No, she wouldn't. But, she asked timidly, would the living room be all right?

"Sure," I replied, swallowing my surprise. The number of people who would put a cat box in the living room was even smaller. But then, not many people have rabbits hopping through the house, either.

So we set up the box in the living room, screened by some very nice potted plants for privacy and resting on some plastic carpet runner for easier cleanup of scattered litter, and we also put plastic in the other areas where the cat had been urinating. I told the owners to leave the cat's play toys on the carpet runner to help him identify these as play areas rather than bathroom areas. (If this doesn't deter the cat, I generally recommend the next most aversive step, covering the plastic with some chlorine-based household cleanser to make it unattractive. Finally, if success is still limited, we turn the carpet runner feet-side up, usually the most aversive method of all. Each of these procedures is usually given a one- to two-week trial.)

The Thompsons had the carpets professionally cleaned, had the rabbits and dogs and cat dipped, and had the house "bombed" with an insecticide to eliminate the fleas, which were rampant and couldn't have been helping the situation. At my suggestion, they stopped punishing Buster and started praising him consistently for using his box.

Buster just loved the new box in the living room. The fact that it was in a huge area with people and pets milling around didn't seem to deter him at all. As I said, he had his own definition of privacy.

The clients were happy to clean out the box once a day; all the dips and cleaning and sprays and medication and praise and lack of punishment seemed to do the trick. Soon, according to the Thompsons, the cat's dilated eyes were back to normal—along with his toileting habits.

« »

Whether it's in the living room or the attic, you do need to think carefully about where to put the box. This is where privacy and escape potential come into play. Again, some cats are more sensitive to these factors than others are, just as some people never close the bathroom door, and others never fail to lock it.

Locating the box properly for the cat's convenience may seem like a common-sense move to most people. You'd be surprised. I remember one client who called me because her previously well-behaved cat was urinating all over the house—everywhere but where the box was located, which was at the bottom of the basement stairs.

When I questioned her to find out what circumstances, if any, had changed, the client revealed that she had taken in five or six dogs temporarily, and that they were housed—you guessed it—in the basement. Whenever the cat walked down the stairs to her box, the dogs set up a terrifying, ferocious racket. So the cat simply stopped going down there. It hadn't occurred to the owner to change the location of the cat's box as circumstances warranted. It was no easy task getting that cat back to normal after this stressful period.

While this cat lived in a combat zone and Buster's environment was like Grand Central Station, Tansy—a five-year-old neutered female—lived in peace and quiet with her newlywed owners. But her need for escape potential was as strong as Buster's, and it was completely overlooked by her owners when it came time to pick a spot for her litter box. Tansy had begun defecating on the floor soon after the honeymoon, and the young marrieds were not pleased.

As I chatted with Richard and Melissa Jordan, I took stock of the setting in which the problem had arisen. They had moved into a beautiful new condominium with a very open floor plan and wide halls and entranceways, bringing the wife's longtime pet with them.

People usually stick the litter box in the least obtrusive place, especially in a beautifully decorated home, and this couple was no exception. I found the pan under an elaborate towel rack, behind a chair in the lavish master bathroom. Although the door was always left open, the box was situated in such a way that the cat couldn't see out of her little area, much less *get* out in a hurry, should anything alarm her. Worse yet—from Tansy's perspective—the box had a cover that further restricted a view of her surroundings.

So Tansy's box was long on privacy but very short on escape potential. All one had to do was look at the location in which she had been defecating for a very clear clue to what was going on in her fluffy white head.

"This is where she has been going," Mrs. Jordan told me, pointing to a spot on the bathroom rug not four feet from where the litter box was located. "If it was at the other end of the condo, I could understand. Then I would know that she forgot where her box is, or she didn't have time to get here, or something logical like that. But to do it practically right on top of her box doesn't make sense!"

"Ah, but it does make sense to Tansy, and from her perspective it's perfectly logical," I answered in defense of the cat. "Please come and stand on this spot where she's been defecating, and look out the bathroom door with me."

Mrs. Jordan walked over, planted herself on the rug where she had found the accidents, and turned to the door.

"If you were Tansy, how far could you see from there?" I prompted.

"Well, I could see out the bathroom, through the walk-in closet, out into the master, then into the hallway to the top of the stairs."

"Do you normally keep all these doors open?"

"Yes, we do."

"OK, now step over to where Tansy's box is located," I directed.

"I can't really get to it without moving the chair," she faltered. "But I guess that her view consists of the inside of the box lid, and the bathroom wall, if she looks out the opening. So what?"

I explained a cat's need for escape potential in order to feel comfortable enough to use a box in any one location.

"Your cat didn't have to go far to find herself a better spot, just a few feet away."

Mrs. Jordan looked at me sheepishly. "I get your point," she said. "But why should she need to escape anyway? Nobody around here chases her. And we don't understand why she stopped using the box all of a sudden."

"That's the trouble with boxes that are poorly located for escape potential," I answered. "The cat can't really tell if there is a threat or not when she hears an unusual noise, because she can't see out. Almost anything might have startled Tansy when she was in the box, and you wouldn't have even noticed. But it could have been enough to cause her to avoid that location and pick a new, safer one. Anything from a dropped shoe to an alarm clock ringing could have scared her."

"I see," said Mrs. Jordan. "It does seem to make sense; Tansy's always been a bit skittish." She thought a moment. "This is all well and good, but I'm not about to move the box to the middle of the bathroom floor," she said firmly.

I glanced around the room. "You don't have to. Let's try it over here, if that would be all right with you." I pointed to another area on the far side of the bathroom that provided a "window" through the bathroom to the walk-in closet and beyond. "This will probably satisfy your cat's need to see what's around, as long as you don't mind discarding the lid of the box."

"Considering the alternative, I'm willing to give it a try."

I gave Mrs. Jordan a few more suggestions aimed at making the box more attractive than the rug, and asked her to call me with news of Tansy's progress. Over the next few weeks we were all pleased to find that a new outlook was just what the doctor ordered, and the newlyweds were soon able to put the cat's troubles behind them and—for better or for worse—concentrate on each other.

« »

Sometimes the owner is doing everything right, but because each cat is different, the solution to the problem may be elusive. Sasquatch, a huge white-and-gray mixed-breed male, had started defecating all over the house for no apparent reason. Every other day, Lydia Berlinger would find another mistake. Making the problem worse was the addition of a five-month-old black male kitten, Critter. Although the two cats got along famously most of the time, Critter had taken to chasing Sasquatch out of the back-hall litter box whenever he could catch the big cat in it. This quite understandably led to four-year-old Sasquatch avoiding the box whenever he knew Critter was in the vicinity.

When I met with Lydia, a solution to the problem seemed obvious. A roommate was moving out in the next few days and would leave the second bedroom unoccupied.

"Your roommate's move is great news for Sasquatch," I told Lydia with some confidence, after taking stock of the situation. "I believe that Sasquatch probably needs some privacy and escape potential, and the empty bedroom is just the place to set up a second litter box."

"I hope you're right," replied the cat's frustrated owner. "By the way, Sasquatch won't even step into the box if it is the slightest bit soiled. I have to clean it out constantly."

"It's good you noticed that," I answered. "Let's try to accommodate the cat's preference. I know it's a nuisance, but a clean box will be much more attractive to him and should help solve the problem more quickly."

Lydia agreed to keep checking the box several times a day, and we made the setting as inviting as possible. We placed the box in a corner away from the door and screened it with plants for privacy. The whole effect was as pristine and attractive as an ad in *Cat Fancy*. When my client called a week later, I expected to hear about at least some degree of progress. To my surprise, there was none at all.

"Sasquatch isn't using the box," Lydia reported bluntly. "I've been keeping it really clean, and Critter doesn't venture into the bedroom at all. But Sasquatch is going all over the bedroom floor."

Back to the drawing board. I reviewed what I knew about the situation. The problem always occurred after the cat's dinner, between five and eight o'clock in the evening. The cat "looked guilty" just before or after defecating outside the box (probably the result of some verbal reprimands), and—most interestingly—if prompted by his owner, he would usually go to the box and use it.

"I know you've spent a lot of time and effort making the box attractive to Sasquatch," I told his owner, "but something about the new location must be wrong for him.

"Let's try moving the box to the hall bathroom," I suggested. "Keep an eye on him after dinner and quietly remind him to go to the box if he seems to be sniffing around or gets that guilty look on his face. And don't rebuke him for accidents."

"OK," Lydia agreed. "I'll call you next week."

A week later I answered her call with some trepidation. But happily, there was definite progress. Sasquatch had adjusted very well to this latest move, and when coaxed into the small bathroom, responded with very little confusion about where to go. The bedroom, though perfectly set up, had apparently offered too much space, and too many alternatives to the box, for this cat to deal with. Within another few weeks the problem was under control to the owner's satisfaction.

This case really brought home to me the fact that there is no one prescription for success that works for every problem and every cat. The personality, preferences, and history of the feline must be taken into account; the first approach may not be the best one. And if a solution doesn't fit, the cat will let you know in no uncertain terms.

Where the litter box is concerned, the important things to remember are these: In general, the litter box should always be placed away from areas the cat associates with other important motivations, such as sleeping, feeding, and playing. The box should be kept clean and should be situated in a location that affords the cat both privacy and escape potential.

Never punish a kitten or cat for missing the box by rubbing his nose in the soiled area, shaking him, squirting him, or placing him in the box when you or the cat is upset. Instead, think about what might have made the box an unattractive place to use, and what makes the soiled place a more attractive location—from Kitty's point of view—and make the necessary changes before you have to call in an animal behaviorist and recarpet the house.

3

The Crime of Punishment

Whjle cats can bring us enormous pleasure, they are also capable of infuriating us with their misbehavior. So we retaliate, our goal being to teach a lesson. Whether the cat is avoiding the litter box, attacking the next-door neighbor, or clawing the furniture, punishment often goes hand in hand with misbehavior. It's the rare owner who fails to reprimand—if not severely punish—a cat that habitually scratches the loveseat or soils the new carpeting.

Faced with mounting frustration when a companion animal repeatedly misbehaves, many owners become flustered and get caught up in punishment and other desperate stopgap measures that will not have a positive effect on the cat's behavior in the long run. Punishing the cat stems from the owners' reasoning that they have to try *something* to let the cat know that what he has done is inappropriate.

When owners call me with problems, I've found that some of them assume that the animal behavior consultant plans to hurt the cat in order to help, because that's what they've been doing. They may have read or been advised to "smack him with a rolled-up newspaper," "shoot her with a squirt gun," "flick his nose with your finger," and the like. Many people think this kind of inhumane treatment is perfectly all right.

At this point I usually ask, "Well, when you hit the animal across the face, or rub his nose in it, does it seem to work? Does it stop the biting [or defecating on the couch, or whatever]?"

"Well, no, it upsets him a lot, and then the problem persists or actually gets worse," they admit, followed by, "but everybody told me . . ." or "but the book said. . . ." I then explain that the more aggression or other misbehavior is met with aggression, the more upset the cat is going to be and the more misbehavior will probably result. So I don't use physical punishment in the treatment program.

That makes owners curious. On one hand, they wonder how I expect to correct the behavior without punishment, and on the other hand, they are relieved that the animal—who is, after all, a member of the family—is not going to have to experience any more punishment and pain.

That physical punishment usually backfires, making the animal more angry or more fearful and aggressive, was never clearer than in the case of Tigger, a one-year-old male cat who had been attacking its owner's legs—and drawing blood—for six months.

With her inappropriate attempts at discipline, Tigger's owner had actually *taught* the cat to attack her. The cat had started out with a litter-box problem as a kitten. When he missed the box, Amy Sullivan, his owner, went for the age-old cure. Tigger's response to the indignity of having his nose shoved into the wet spot was to bite her.

Then the battle escalated. When the cat bit her, Amy yelled at Tigger and rapped him on the nose with her knuckles. She was very consistent, rapping his nose each and every time he retaliated. This punishment had only one effect: Now, every time Amy started to present her knuckles—even when she was just planning to pick up the cat or feed or pet him—he'd lunge at her ankles. Clearly, the punishment had become the signal to attack. Eventually Tigger would do so often, even if Amy's hands weren't involved.

When Tigger did attack her, Amy kicked him off her leg. This resulted in more fear and pain, and the cat would understandably run away and hide. So after six months, Amy had a cat that would do one of two things: run away from her or attack her. What she really wanted, she told me sadly, was one she could pet and enjoy.

"It's gotten so bad that I have to pick up Tigger and carry him around as soon as I come in the front door so that he won't attack me!" she exclaimed. "Do other people go through this?"

I assured her that she was not alone, and that there were basic changes we could introduce in how she interacted with Tigger that would make both parties happier. While I sympathized with my client, it was readily apparent that she had more or less dug her own grave. As she came to this realization, we both turned our sympathies toward the "naughty" cat.

Tigger's initial encounter with Amy's knuckles resulted in "fear-induced" (or defensive) aggression, and it was clearly a case where physical abuse had severely aggravated what had started as a rather minor and occasional problem. One of the ways I could tell that this cat was fairly reactive—that is, was responding emotionally to a signal or stimulus—was that his eyes would dilate just before the attack. Dilated pupils may signal that cats are overly aroused and probably about to do something regrettable, such as losing it on a rug or attacking somebody. (In a predatory attack, the eyes generally don't dilate as noticeably.) Tigger's eyes became quite dilated before he pounced, and with the history of physical punishment, it was understandable.

Some of the more predictable attacks took place at feedings. I watched Amy put the food bowl down, then walk past it as she did at every feeding. Tigger instantly jumped at her ankle, as *he* did at every feeding. The stimulus was so cued in, as in a ritualized pattern, that he attacked her more or less automatically.

To stop this problem, Amy and Tigger would have to break their daily pattern. I told Amy to back away from the food bowl instead of walking past it, so the cat missed his cue and concentrated on his food. She tried it and reported that it worked right away. That particular problem behavior was halted easily and instantly by a simple change of routine.

To take care of the rest of the attack problems, I instructed Amy to stop punishing the cat and play with him instead, tossing toys away from her and from him. In order to reduce the cat's fear, Tigger needed to learn that Amy's presence signaled play, not punishment.

"As you walk by," I said, "Tigger should be playing with something instead of racing over to bite you on the legs. Use his eyes as a barometer of impending aggression. If he follows you, try throwing another toy—a crumpled piece of aluminum foil, a stuffed sock on which to redirect his biting. Use a different door to come in and go out a couple times each day, for a week to break the 'I enter, you bite' ritual. Keep backing away from the food dish after you feed him. And number one, again, don't punish the cat anymore."

I presented her with a program for solving the cat's litter-box problem and left, feeling that things would probably improve rather rapidly if Amy could adapt to a new style of handling misbehavior.

Happily, Amy rose to the occasion. She stopped yelling and rapping and putting Tigger's nose in defecation. After a few weeks, Tigger was amusing himself with his toys, using the litter box more consistently, and showing his owner those big, dilated pupils far less frequently. So this was a treatment program that hinged on the simple basics of humane treatment and a redirection of the cat's energies towards playthings and away from retaliation.

Whatever the situation I encounter as I take on clients, in no case do I ever blame the cat for the behavior problem, nor do I blame the owner, although I may privately be upset about what the owner has inadvertently put the cat through. It's very rare, in my experience, that an owner who calls me for help has deliberately abused the animal or somehow intentionally caused the behavior problem to occur. Sometimes I find that I really have to sympathize with both the cat and the owner but at the same time let the owner know that I'm not blaming him or her for the problem.

Guilt trips have no place in animal behaviorist–client relationships. Clients made to feel guilty are less likely to be open to suggested solutions and less likely to participate in programs that could help their cats. They may be so worried about what a "professional" thinks of them or what they might have done to head off the problem or what they think they did to cause it, that they are not able to pay attention to the details of the treatment programs or concentrate on the theories or concepts behind them. Even if the problems result from things the owners might have done wrong inadvertently, it's best not to dwell on blame. I try to use the circumstances just to help clarify diagnoses and treatments.

My determination not to place blame was tested when I went to work on behalf of unfortunate Champ, a two-year-old male cat I was called upon to help several years ago.

Champ's owner, a young woman named Kathy, had made a lot of mistakes by the time she called me. First, she had spent months trying to train the cat to use the toilet instead of the litter box. While toilet training occasionally works, messing with the cat's established preference for a litter box—usually established early—can make the training sessions stressful and ultimately unproductive, especially if the owner lacks the skill or patience to follow directions and be consistent.

Champ failed to learn how to use the toilet, much to his owner's disappointment, but Kathy's ire was aroused when the cat started to reject the litter box, too. He began urinating in a corner of the living room and on Kathy's bed about three weeks before I became involved in the case.

Caught in the act, the hapless cat was yelled at and punished.

Kathy's method of discipline was to rub the cat's nose in the urine, or to put several drops of vinegar on the cat's nose, or both.

It made me wince to hear her describe these episodes. When I suggested that it was probably not a good idea to do those things and pointed out the sensitivity of the cat's nose, Kathy was genuinely surprised. She had no idea that her methods of discipline could be thought of as abusive to her pet.

"I just thought I should let him know he shouldn't go outside the litter box," she said with a shrug. "And it seemed to work—for a while, anyway."

"I can see why this kind of discipline would make your cat avoid the bed," I commented. "But physical punishment like that is just a stopgap method; it doesn't really change the cat's preferences. Let's see if we can come up with a better approach to helping Champ with his problems."

Complicating matters was the fact that the cat was covered with fleas and had suffered the indignity of being treated with a foam product that had made him ill. Since Kathy hadn't had the house "bombed" at the same time, the fleas were a recurring problem. An electronic flea collar hadn't helped, either. So Kathy had simply given up on the fleas.

Finally, as a last-ditch effort to stop the cat from urinating outside the box, Kathy had moved the cat and its pan, along with Champ's food bowl and sleeping blanket, to the tiny bathroom, assuming that if Champ were confined to such a small space, he couldn't possibly have an accident. At least he wasn't going on the rug or bed while he was in the bathroom.

Now, under certain circumstances, confining a cat to the bathroom for a short period can help him refocus on his litter box. The ideal conditions call for a large bathroom with enough space to widely separate the food, the litter box, and the cat's resting area. There should be a window with a sill the cat can reach to sit on, some sunshine for basking in, and—for snoozing—one or two inviting elevated areas (not counting the counters around the sink, which are too cold and hard for most cats). Without these basic requirements, the cat simply doesn't have enough good space around which to comfortably organize his daily activities.

Even at best, with the perfect bathroom, the cat's learning curve will most likely go flat after two or three days of confinement. Four days is really stretching it.

Well, Kathy kept Champ in her tiny, windowless bathroom for two weeks. And when she discovered an accident—usually long after the fact—she rubbed his nose in it, believing she was doing the right thing.

There was no point in dwelling on what Kathy did or why she did it. Many people have simply been taught that physical punishment is the correct way to deal with wayward pets, just as some parents believe that sparing the rod will spoil the child when kids get out of hand.

Needless to say, by the time I got to this cat and its well-meaning owner, I had my work cut out for me. Before I could even begin to concentrate on a behavior program for inappropriate elimination, we had to focus on stopping all the abuse and folk remedies, getting rid of the fleas, and helping to build up a good relationship between the cat and Kathy. Perhaps because of all the cat had been through, Champ never became the "perfect" pet. But he and his owner were able to coexist in a much more comfortable relationship because of what Kathy learned during the six-week program.

It's cases like this where I regret being the last resort, for the cat's sake as well as for the owner's. The proper early intervention can often save both of them a lot of misery.

Once they learn about effective alternatives to punishment, people who have thought enough of their pet to call for help are uniformly happy to try more humane ways to alter their companion animal's misbehavior.

Although harsh physical discipline may work to interrupt a

misbehavior, its long-term consequences and side effects may be costly to both cat and owner. Consider a physical prompt or correction as an alternative if you feel you *must* discipline your cat. Unless it is extremely mild or gentle, physical punishment does not work and is not appropriate. Just a little *touch* of the finger on the nose or the chin is usually enough to displace the mouth away from one's hand or finger if a cat starts showing a biting problem. Similar to what a mother cat might do to correct her kitten, a soft tap is understandable to the cat, and it's something an owner can consider doing, along with more positive options to help resolve the problem.

Certainly don't spank a cat with a newspaper, or jerk it by its collar, or rub its nose in urine. Of course I don't recommend any of those measures for dogs, either, but people are more prone to try them with dogs than with cats.

If your cat starts to show a behavior problem, sometimes just a voice—"Ah, ah, ah!"—is enough to interrupt the activity. When the cat looks at you, you can say, "Gooooood Kitty," in a sweet, smooth voice. Then your cat starts to learn the difference between doing something that gets a harsh "Ah! ah! ah!" and doing something (anything that is different from the problem behavior) that gets a continuous, smooth sound.

The idea is to reduce the misbehavior by interrupting it or redirecting it to an acceptable outlet, not to punish the pet. You want the cat to stop what it is doing, to do something else, and to get praise for the preferred behavior.

We change behavior by redirecting it, or by inducing in the cat a happy or calm mood—or, in cases of aggression, a mood incompatible with the fear or pain, or whatever else motivated problem behavior. And once you change the mood, then changing the behavior is easier, for the cat and the owner.

The goal of my work is to help people enable their cats to do things differently. First I teach the owners to indicate to the animal (either directly or by restructuring the living environment) that *this* is desirable and *that* is not desirable. It's a simple discrimination learning task. Then we go on to create a framework within which the cat learns something to do in place of the undesirable behavior. Believe me, it works a lot better than physical punishment, and leaves owners and their cats much happier with themselves and with each other.

4

Eating Habits
Weird and
Wonderful

Today, cats have an overwhelming array of food products to chose from, beginning with kitten food and going all the way to senior citizens' fare, with every gourmet delicacy and kitty treat imaginable in between. Special diet fare is available for cats allergic to regular cat food. A few cats eat better than their owners, living solely on lamb, liver, and chicken lovingly prepared by their adoptive parents (not necessarily to the benefit of the animals' health). Commercial cat food is the main reason—along with regular veterinary care—that cats today live a lot longer than did their ancestors, who were fed on tasty but nutritionally inappropriate table scraps and seldom inoculated against any diseases.

As long as the daily intake is well balanced and nutritionally sound—you can check with the vet on this point—what the cat eats is pretty much up to your cat's taste and your pocketbook. A cat that doesn't like the menu will notify you (sniffing at the bowl of food and then walking away with a shake of the leg is a clue), and then you can try something else.

The most important thing to know about a cat's behavior relating to food is that *cats are naturally nibblers*. Active cats prefer to eat somewhere between twelve and twenty very small portions a day. They are diurnal eaters, which means they eat both during the day and the night.

This is why they prefer to be fed "ad lib" rather than only at fixed mealtimes two or three times a day with the food removed. Your kitten and cat should have food that won't spoil and fresh water available at all times. Knowing this, you'll no longer consider your feline finicky if he takes a few bites and leaves the rest in the bowl. The cat will be back.

If you ignore your cat's natural eating preferences and take the bowl away after a few minutes, you run the risk of allowing a bad habit to develop. The cat may learn that the only way to get enough food is to race up to the bowl when you set it down, gobble up as much as possible before you snatch it away, and—because this is not the eating pattern for which the feline digestive system was designed— promptly regurgitate it onto the white broadloom. Or worse.

One of the most interesting cases I've dealt with involved a cat bingeing and bolting her food, but it took me quite a while to make a connection between her eating habits and the ensuing behavior problem, which was much more serious than vomiting on the rug.

"My cat Marmalade has wrecked thousands of dollars' worth of carpet, and I've just had it all replaced," my new client, Laura Hunt, explained as she showed me into the living room of her small townhouse early one morning. "For the past three weeks she's been going on the *new* carpet," she went on. "I'm about at my wit's end."

I caught a glimpse of Marmalade's orange tail as she scooted up the staircase. The former stray had changed from a nice, affectionate pet to a nuisance that kept her owner, a school nurse, awake at night wondering what to do.

As Ms. Hunt went on to describe the problem, it appeared at first to be a typical case of rejecting the litter box and marking with urine.

The cat was urinating or spraying all around the house: along the baseboards in the living room, on the speakers attached to the entertainment center, upstairs in the guest bedroom, in corners along the wall by a bookcase, along the walls near the front door, and so on. The problem had been going on for more than a year, Marmalade's owner told me wearily. She'd put up with it until the cat started soiling the new carpeting. Now it was panic time.

I reviewed with my client some of the ways to make the cat's litter boxes more attractive and the carpeting less so. We decided to move one of the boxes to an opposite corner of the living room, behind a large plant. It was an area that the cat was already soiling, and it offered more privacy and escape potential than the old spot. A second box was in the upstairs bathroom. I went over the use of plastic carpet runners on high-risk areas throughout the house.

"I tried keeping Marmalade in the bathroom while I was gone during the day," Ms. Hunt explained. "She used the box. But she also chewed up the doorjamb and knocked over all my toiletries. I just couldn't keep her in there another day."

"Sounds like you did the right thing," I said. "Not all cats can handle confinement in a small area."

"She can be pretty hyper," the owner added. "She zooms around when the doorbell rings and doesn't like catsitters or strangers."

"So I noticed when I came in."

"On the other hand," she went on, "if I'm getting ready to go on a trip, she crawls under the bed when she sees my suitcase." At least the cat didn't urinate in the bag, as had several cats I'd met.

As we talked, Marmalade ventured back down the stairs and gingerly walked over to the living room couch (carefully picking her way past former target areas) to give me a curious look. Her pupils, I noted, were the ordinary slits of the nonaroused cat. This one didn't appear to be particularly stressed out, and the zooming sounded pretty typical.

Ms. Hunt emphasized that—unlike many cats that urinate

outside the box—Marmalade routinely used her box as well as the carpeting for elimination. So I knew that the cat wasn't avoiding the box; this could probably rule out anything traumatic that had happened there or a medical connection. We already knew that the cat had a clean bill of health from the vet.

While active and intense, the cat didn't seem to me to be anxious and suffering from an overload of stress, which can also cause litter-box problems. And with no outside or other indoor cats to contend with, I was fairly sure the cat was not marking her territory—which would have been unusual behavior to expect from most females anyway. Although I had just recommended some changes regarding Marmalade's litter boxes, nothing about the way they had been set up was bad enough to cause the cat to reject them altogether. And the new carpeting lacked the familiar scent that previously drew the cat back for more accidents. So far it all added up to very little in the way of reasonable explanations for the behavior problem. There seemed to be more to this problem than I was able yet to grasp; I clearly needed to keep probing.

"Where does Marmalade sleep?" I asked Ms. Hunt.

"On the bed, with me," she answered. "And she doesn't use that area as a litter box, thank God."

"That's a good example of competing motivations working to your advantage. Most cats are very reluctant to soil the areas where they eat or sleep."

"Well, she definitely wouldn't let anything interfere with her eating," Ms. Hunt commented with a quick frown. Knowing that many cats barely flirt with food, I found my interest piqued by the remark and asked Ms. Hunt to tell me more about Marmalade's eating habits.

My client was more than happy to do that. It seemed that Marmalade was a "binge eater."

"If she had the chance, I think she'd probably eat and eat and eat until her belly was dragging on the ground," she commented. "The more I give her, the more she gobbles up. So I stopped refilling her bowl a long time ago—she'd weigh a ton if I hadn't."

"And what is her feeding schedule?"

"One serving in the morning before I go to work, about 6:30 A.M., and another at dinner time—six o'clock at night," she replied. "And Marmalade always cleans her bowl in two minutes."

I leaned forward. "You don't leave her food at night, or while you're gone all day?"

"Heavens, no! The cat would be obese." Noting my skeptical look, she folded her arms and added, "Dr. Husak, who referred me to you, didn't think she'd get fat either. But I'd say she has a real bingeing problem, and that's why I feed her twice a day."

"I'll grant you that taking away the food has kept Marmalade pretty trim," I acknowledged. It wasn't surprising that Marmalade bolted her food. With twelve hours between meals, the cat had to be famished.

"She's not overeating, but she is urinating all over the house," I remarked.

Now it was Ms. Hunt's turn for a skeptical look.

"Are you saying that the urination problem has something to do with her bingeing?" she asked.

Answering that it was certainly possible, I suggested we talk for a moment about how cats typically eat. I told her that cats were notorious nibblers, going to their dishes many times during the day and night for small snacks. This is the eating that seems to be the most appropriate feeding method to keep felines happy and healthy.

"I know that most people leave food out all the time for their cats," Ms. Hunt said. "But I can't do that for Marmalade. She'll eat too much and get too fat."

"But if Marmalade were eating more frequently, she'd most likely stop bolting the food," I said. "She's probably very hungry with only two feedings a day."

"I just couldn't take a chance," Ms. Hunt said firmly.

I looked at her in exasperation. Neither she nor her cat showed any signs of overweight; if anything, they were both on the thin side. But, the woman was a dietician. She was extremely aware of calories. For all I knew, she had just lost fifty pounds and projected her battle of the bulge onto the hapless cat.

I saw it was pointless to press further for ad-lib eating. Swallowing my frustration, I decided to try another option. I had remembered a tip from a colleague, Amy Marder, who successfully faced the challenge of finding a high-bulk, low-calorie food for some of her clients.

"I think Marmalade might be feeling very uncomfortable as the hours pass without anything in her stomach," I went on stubbornly. "This may be causing her to urinate wherever she happens to be when the hunger pangs make her irritable or anxious."

"Do you really think so?" my client asked.

"I think it's something we might want to consider," I answered. "Would you agree to add to Marmalade's breakfast and dinner a nonfattening food that will give her a more satiated feeling during the hours she's not eating?"

"Such as?"

"Such as some romaine lettuce," I replied.

Her eyebrows shot up. "Lettuce?"

"That's right," I answered. "There's a fifty-fifty chance she'll eat it; cats seem to either love it or hate it. But it's a high-bulk, low—"

"I know that," she cut in, smiling. "Well . . . I guess it couldn't hurt."

I asked Ms. Hunt to add a cut-up portion of romaine lettuce to the cat's food at the breakfast feeding, and told her that I would touch base with her veterinarian to discuss other possible changes in diet. I wanted to find out if there was any high-bulk, low-calorie cat food on the market that might be appropriate in this case. I knew that dog owners had no trouble finding such a diet, but I didn't recall ever seeing that kind of food for cats.

I called Dr. Husak, who was glad to hear Ms. Hunt and I were trying to resolve Marmalade's problem, and he said that there was one good cat food sold in pet stores that met the requirement of being low fat and high bulk. With the go-ahead from the vet, I recommended that Ms. Hunt switch to the new cat food and lettuce regimen twice a day and phone me in seven days. She agreed to give it a try.

A week later she called with great news: Marmalade loved the new food, she loved the lettuce, she loved the new location of the box, and there hadn't been any mistakes since they started the program.

"The only thing is," she stopped for breath, "Marmalade has gained a half pound, and I'm worried about it."

"Well, let me discuss that with the vet," I suggested. "Please keep doing what you're doing."

When I raised the weight gain with Dr. Husak, he told me the cat could stand to gain another half a pound, in his opinion, and there was no reason to worry about her weight.

The following week there were still no mistakes; a remarkable record. But Ms. Hunt sounded discouraged. "I found scratch marks on the carpeting upstairs, where Marmalade used to go," she said. Although she didn't see any signs of urination, she was afraid the program wasn't working.

I had to bite my tongue. I thought of the hordes of cat owners who would have given their right arm for results such as those Ms. Hunt had achieved.

"Really, you worry too much," I chided her. "Marmalade's doing fine. Enjoy her."

"Do you think so?"

"Well, you tell me," I replied. "Has she made any errors since we started the program?"

"No," she admitted. We left it at that.

Whether their favorite food is chipmunks or Brie cheese, I must emphasize that cats are not normally finicky eaters. This is another myth, probably foisted upon cat owners by dog people, who are used to pets that routinely wolf down four cups of dog chow in a single gulp, finishing up with anything they can steal from the garbage can for dessert. By comparison, cats do seem downright dainty as they nibble at their dinner.

Although there are differences in eating preferences among cats, I'll go out on a limb: Finicky felines are made, not born. Offered only canned salmon and cream since leaving the mother, your kitty has been trained to refuse anything else occasionally offered.

Some studies indicate that kittens exposed to a variety of three or more foods during the first eight weeks are more likely to prefer a variety later on. Kittens like trying new foods—but will continue eating them only if they really like the flavor after a couple of different meals.

By eleven weeks of age, they typically become more rigid about what they will accept. If they've tried only one food, eventually they'll be just like the eight-year-old kid whose entire repertoire consists of grilled cheese, tuna, and bologna sandwiches because that's all he got when he was three.

Conversely, I've known kindergartners who eat oysters and squid because they were introduced to a variety of foods as toddlers and don't know any better, and cats that eat everything from vegetables to popcorn. Just check with the vet if your feline's food preferences seem questionable or if you think the cat is too thin or too heavy.

If your cat's appetite is good but includes eating crazy things that aren't even food, you may have a difficult-to-solve problem requiring an animal behaviorist's help. I'm not talking about eating plants and grass, which is common in cats and generally harmless (as long as the plants are not poisonous).

The strange ingestive behavior I'm referring to is known as "pica." The cat, for no apparent reason, seems to crave foreign materials such as wool or other fabrics, wires, rubber, dirt, and the like.

I've seen fabric eating most often in Siamese and Burmese cats and occasionally in other oriental strains. Woolens, cottons, and synthetics are sought out regularly or sporadically, the way other cats search out sardines. This is among the most difficult cat behavior problems to solve, as I found when I called upon two distraught women, Lynn Terhune and Alexa Quinn.

Waitress and telephone operator roommates with a gray kitten named Walker, they were shocked when I told them their little alley cat might be misbehaving because he was stressed out. The two women were somewhat oblivious to the frenetic and ever-changing nature of their life-style. But by the time we had talked for ten or fifteen minutes, we'd come up with the following list of events that had unfolded during the previous two months.

1. At seven months of age, Walker is neutered.
2. Alexa has new man in her life who visits frequently.
3. Lynn changes to the overnight shift at the diner.
4. Walker acquires fleas.
5. Owners adopt a stray, but take it to the shelter when Walker seems to hate it.

"So you think this list explains everything?" Lynn asked dubiously. I could see why she was skeptical. The "everything" in this case was a rather unusual reaction to a stressful environment.

Walker was missing the cat box frequently, but in his case that was only the beginning. He had become, as Alexa dramatically described it, "an unstoppable chewing machine!"

And his items of choice were, unfortunately, the owners' clothing. When the women were asleep or at work, Walker opened the closet doors, stepped inside, and selected their favorite sweaters, party dresses, and Lycra exercise wear. He would pull the garment off the hanger with paws or teeth, or get into one of the many cubbyholes and bins in the closets, and proceed to chew up the contents.

At first he went to Alexa's closet exclusively. Now, in the past two weeks, he had started venturing into Lynn's. "I guess it's because Lynn wears mostly polyester. I have more hundred-percent-cotton things," Alexa explained proudly.

"You mean all those tacky lace gowns you have," her roommate rejoined.

"Tacky?" Lynn squawked. "What do you call that purple leather skirt the kitty chewed all to bits? I've never seen—"

"So," I broke in, before they could come to blows, "shall I note down that Walker seems to prefer clothing made of certain materials, such as cotton. . . ."

"Lace!" Alexa broke in.

"Don't forget leather," Lynn added crossly.

"And the cat chews and sucks on these things only when you're asleep or out of the house?"

"Heck, no!" Alexa replied. "When we walk in, the first thing he does is run to the closet and try to get something to chew up. Right in front of us. Or he might just pick something up off the bed or the chair. But of course we stop him from doing it if we're around."

We took the requisite tour of the house. There were clothes strewn about the bedroom, I noticed when the roommates showed me the scene of the crime. Their two large walk-in closets had folding doors that the cat could easily push open.

I asked the obvious. "Have you tried keeping the doors shut with some kind of lock or something?"

Lynn gave me a look. "You try opening a combination lock when you're coming home from work in the middle of the night!" she exclaimed. "We can't be bothered with locks. Walker shouldn't be eating our clothes."

"It's hard to keep things picked up, too, with our schedules and all," Alexa added, throwing an armful of clothing into a laundry hamper in the corner of the bedroom.

"I can imagine," I said. "Well, there are a couple of ways to deal with Walker's problem. First, do you think you can deflea him? He's probably very uncomfortable and it's possible that chewing and sucking must give him some kind of relief or comfort." I surmised that the chewing and sucking might be releasing endorphins in the brain, a natural opiate that masks painful stimulation—or in this case, the discomfort of flea bites.

The two women agreed to do that, and I went on to explain the rest of the program.

"I don't want you to physically punish Walker, but if we can startle him enough to make the closets seem an unpredictable, unpleasant place to be, he will stop going in them," I said.

"My boss at the diner said I should squirt him with a squirt gun," Lynn offered.

"And did you try that?"

"No, I didn't," she replied. "I didn't want to get my clothes all wet."

"Well, that's one reason, but there are other considerations." I explained that to squirt a cat with water could add to its stress level. When I first started doing house calls, some people were recommending squirt guns, but clients found that the squirted cat began to consider the owner pretty scary and someone to be avoided, which is not what most people want to happen with their companion animals.

"If you use a remote punisher, such as the upside-down mousetrap people put on their counters to discourage cats from jumping, the cat is startled momentarily when the mousetrap goes off, and it doesn't look as if you had anything to do with it," I explained.

"Or you could go to an electronics store and get a little laser alarm system that makes a lot of noise if anyone or anything breaks through the beam of light." I pointed at little Walker, who was walking back and forth eyeing Alexa's closet. "This again will seem to have nothing to do with you, and Walker will learn pretty quickly not to go near the closet."

"I'd love to hear that thing going off when I've just fallen asleep at 5 A.M.," Lynn said dryly.

"I don't think we could afford it," added Alexa.

Back to the drawing board. If a client isn't comfortable with one suggestion, or if it doesn't work, several backups need to be considered. Another way to approach the problem was not to try to force the cat to stop chewing but to give him some alternative items to chew on, possibly the items he already preferred.

"Oh, sure, but he prefers my clothes!" replied one of the owners.

"I just want him to behave like a normal cat and stop eating my sweaters," said the other.

"Let's take the approach that it's OK to chew, as long as the item isn't your clothing but something a cat normally plays with," I said. "We don't have to try to force Walker to stop chewing. Will you give him some pet-store cat toys that he can chew to his heart's content?" I asked. They nodded.

"These won't resemble in any way the kind of items Walker preferred in the past," I advised them. "This way it will be easier for him to discriminate between inappropriate and appropriate items. The items will be made of hard rubber or something other than cloth."

"We can do that," Lynn agreed.

"Also," I added, "it might be helpful to give Walker a feeling of having a very full stomach, so he relies less on taking something into his mouth to satisfy himself. There are canned high-bulk cat foods put out by—"

"No way we can afford canned," Alexa cut in.

"OK, then try feeding him some romaine lettuce, which should accomplish the same thing," I said.

"What about his leash?" Lynn asked. She had told me that she enjoyed taking the cat for walks using a small fabric harness and leash, but that Walker had chewed up his leash.

"Just buy a new one made of chain," I said. "The important

thing is to get the walks going again, since they seemed to soothe him, and they'll be a part of a predictable routine if you walk him regularly."

I also recommended putting the cat on a temporary regimen of a mood-elevating antidepressant, which I would ask Walker's veterinarian to prescribe. This would make him feel calmer as he gave up chewing on his favorite items, I explained.

I gave my clients some instructions for a successful litter-box program and asked them to check back with me once a week.

At the first feedback session, Lynn and Alexa said they had devoted a lot of time to walking their pet and reported that the medication had started to calm him. By the second week, Walker was using his litter box more consistently and showing much less interest in eating clothing. They were pretty happy by the fourth phone call, having tried all the suggestions that they felt were doable. Eventually the cat was down to a maintenance dose of the antidepressant, monitored regularly by the veterinarian.

"Guess what?" Lynn said excitedly over the phone the last time we talked.

"What?"

"I accidently left a brand-new cotton T-shirt on the bed when I went off to work last night. And when I came home Walker was sitting on it playing with his catnip toy—and he hadn't chewed it up!"

"You must be doing something right," I congratulated her. And thus ended The Case of the Feline Fabric Fetish.

Ingestive problems like this one are tough, because I am not sure what's causing them. I know that if we can lower the stress, keep the cat from becoming too hungry, and make whatever the cat is eating taste bad or be unattractive in some other way, we can reduce the problem; but somehow these cats still seem to have the urge to eat nonnutritive stuff such as wool or cotton. That there also appears to be a genetic influence here makes it a little more difficult to resolve.

But if you don't provide the potential wool eater with the opportunity to eat this sort of thing early on, even though there might be a proclivity for it, there is a good chance the cat will get used to and prefer nutritive things that are better suited for growth and good health, physically and behaviorally. So start shelving your sweaters if you suspect your cat is eyeing those soft materials for reasons other than to curl up and take a snooze.

5

Can You Teach a Cat Good Behavior?

" Say 'no' to a cat and he just doesn't get it," argues the dog booster.

"Oh yeah?" the cat lover counters, "Well if your dog is so smart, how come you can't even go away and let him take care of himself for a weekend?" And so it goes. But even in households devoted to both the cat and the dog, you're likely to hear a lot of debate about which species is "better," or "smarter."

Since I've worked and lived with both cats and dogs, I'm sometimes called upon by clients to settle this argument. As a big fan of both species, I have to respond the same way many of my colleagues do when faced with this query: Cats are smart for cats, and dogs are smart for dogs. And because dogs and cats do different things, it's not fair to compare them.

Is that ducking the issue? Not really. I readily acknowledge that cats will often just ignore the owners who tell them "no," whereas dogs will respond with some indication of having heard and understood what the owners are telling them. But I believe that's not so much intelligence as obedience or compliance growing out of each species' level of dependence on human companions.

It may be that it is not as important for cats to obey in order to get what they need as it is for dogs. After all, dogs must rely more on others to satisfy their biological needs, and thus appear in general to be more receptive to human communications such as words, gestures, and expressions. So some might say that in refusing to acknowledge commands, because they don't need to, cats are actually demonstrating they are brighter than dogs. The same argument could be applied to dogs' superiority in "knowing" they are dependent on humans for their needs.

We don't have cats in our lives to show us they are able to sit, stay, lie down, heel, or learn to shake hands and roll over on command. Nor do we look to them for protection. Most cat owners choose felines for companionship and because they like the softness, beauty, or independent demeanor of cats, even though cats are actually quite social.

In fact, there is probably a very similar range of sociability in the attachment of cats to people as in that of dogs to people. But because of their different natures and social structure, dogs will welcome people along on a walk and treat them as part of the family unit, whereas most cats do not expect to come back from a hunt and take a walk around the block with their owners. After all, cats are solitary hunters, which is the predominant reason for the myth that cats are solitary as a species—which just isn't so.

Cats are also a relatively ritualistic species that do well in an ordered and predictable environment, which describes the domestic household scene fairly well. (People with totally chaotic, unpredictable life-styles usually choose not to keep pets or are more likely to have cats and dogs with behavior problems stemming from the instability encountered in their daily lives.)

To learn to deal with life-threatening situations they may encounter, cats need to practice how to escape and how to do battle. They learn this (including how to inhibit their bites) from their mothers and littermates. They also learn how to catch their prey during both day and night, aided by sensory apparatus that includes keen eyes and ears they can rotate for better hearing.

There are probably as many individual differences among kittens in a litter as there are differences among litters of cats. In other words, there is a tremendous individuality in cats. Some are very timid, some are much more outgoing; and if they're outgoing, they're going to explore more and learn more.

But cats that are timid or neophobic (avoiding new or novel objects, people, sounds, etc.) are less likely to explore, and they are more likely to stay in one location, so they are likely to learn less. I don't think it's a direct measurement of how intelligent they are. It has more to do with the kind of temperament that best allows them to learn.

Although cats' temperaments, including their tendency to explore, react, or comply, may affect what they learn from exposure to other cats, dogs, people, objects, and experiences that have pleasant or unpleasant consequences, temperament and intelligence are different constructs. Two different cats with identical temperaments or personalities may take more or fewer attempts to figure out that when the cupboard door opens, food will be placed in their dish. But for the companion feline, this is probably irrelevant.

Well-cared-for cats don't have to figure out where their next meal is coming from, how to stay warm in the winter or cool in the summer, or how to defend other desirable resources to best ensure breeding success. In the domestic animal, all that is taken care of.

Although I enjoy watching my cats Turk and Domino figure out how to get me out of bed in the morning, I don't believe those behaviors constitute the same concept of intelligence as we use for humans. Because its practical use in describing feline behavior is limited, perhaps we should avoid the term altogether. A more fruitful use of our curiosity over understanding our companions' behavior may be to bear in mind what we've observed about our own cats' temperaments, personalities, and ability to learn.

Cats are definitely trainable, if the number of books claiming you can teach your cat dozens of tricks are any indication. People who train cats for a living find that it's not too difficult, if they start at the right time with the right goals in mind. You probably don't want to have your cat try to be doglike any more than you would want your dog to be catlike. But you might want to teach the cat to climb on top of the refrigerator to eat or to play with certain toys and leave other things alone or to go through a cat door to get outside. These sorts of tasks are fairly easily taught to ordinary cats.

Can the cat be taught not to attack a pet bird? Yes, sometimes, if you use "aversive" tactics. You can get a realistic-looking feathered bird toy and lace it with pepper or put a little audio speaker on or near it and turn up the speaker to startle the cat when it comes by. Outdoors, if you don't like your cat to hunt, the age-old bell-on-the-collar routine has proved successful. You can also keep the cat inside and try to limit the number of smaller animals who share living quarters—for they are all fair game.

Training cats is a matter of shaping their behavior so they do the things they automatically want to do or of using their natural preferences to teach them to do other things that you want them to do. Just start early if possible; although you can teach an old cat new tricks, it's not as easy. I have a file full of cases to prove it.

Suppose, for instance, your goal is to get a kitten to stand up on her hind legs and reach for a bell upon command. A kitten will love to play and reach, so if you start by holding the bell a little bit over the kitten's head, she will reach for the object almost automatically. As you raise the bell a little higher and then higher, the kitten naturally still wants to play with the object, so she continues to reach.

If you say the word "high" each time you hold the bell above the kitten's head as it reaches up, the kitten will become conditioned to the word "high" meaning "stretch up high to get the bell." Eventually all you have to do is say the word "high," and the kitten will stretch up, ring the bell, and make you look like Gunther Gebel-Williams.

But let's face it, training a cat to ring a bell for you probably isn't a high priority for most cat owners. Do you think you could teach your cat something useful such as going to the litter box on command? I've encountered several owners who have taught their cats just that, and it really came in handy when behavior problems reared their ugly heads.

Chantilly, a seven-year-old female Persian, needed some retraining to know where it was acceptable to go to the bathroom. When I met her, she was not using her box at all, but was doing her business in random locations all over the house.

But like several cats I've become acquainted with in my practice, Chantilly had a leg up, so to speak, on other cats with litter-box problems: If her owner told her to go to the box, she would usually do so. Although it is not generally acknowledged that cats can be trained to eliminate upon command, I have seen enough cases to accept this claim when I hear it from an owner.

Whether the majority of cats have the potential to be taught to go on command is an open question. I suspect that if they were all so easily persuaded, the train-your-cat-to-use-the-toilet routine would be quite a bit more widespread, and cat elimination problems would be fewer. On the other hand, how many people have even thought of trying to tell their cat when to use its box?

At any rate, Chantilly wandered around the house, going all over the place—on a couch or carpet, in the hallway, on the baseboards, and so forth. She did have a cat friend at home, another Persian that she religiously kept from using the litter box whenever possible by scaring it away. (Not that Chantilly would put a paw in the litter box herself.) If she wasn't going to use it, then the other cat, China, wasn't either. Luckily, China was unflappable and managed to use the litter box when Chantilly wasn't looking.

When I met with Jerry and Emily Gardner, the owners, I first went over the most obvious problem areas and made some suggestions for improvement. The food was too close to the litter—Chantilly may have been averse to eating and eliminating in the same general area—so we moved her bowls to a new location.

We put some plastic carpet runner under the box, which was located in a carpeted dinette area off the kitchen. I use carpet runners quite a bit in my work as an aid in helping cats discriminate between appropriate and inappropriate places to relieve themselves. Often cats will approach litter boxes but go instead on carpets because they like the feel so much. By using carpet runners, we give cats a chance to feel the cool, hard texture before stepping into the soft, easy-to-manipulate kitty litter. (As I mentioned earlier, plastic garbage bags don't work as well as carpet runners because, like litter, they can be pushed aside by cats.)

I had the Gardners stop all punishment and be sure to ask Chantilly to go to her box when they saw her sniffing around. (If the owners even walked to the box with Chantilly, the cat was also likely to get in.) They were to praise her after she stepped into the box, and of course, not discipline her for making mistakes—understandable though it was to want to take out their frustration on the cat.

The goal of any kind of litter-box training is to make the box as attractive as possible to the cat, while at the same time making the "preferred," non-litter-box areas as unattractive as possible. To this end, I persuaded the Gardners to put plastic carpet runners in other locations in the house and to wipe them with a chlorine-based bathroom cleaner that works well to keep cats away. I told the owners to use a variety of products to clean up any accidents they discovered, so that the cat didn't come to identify a specific cleaning substance's odor with a place to eliminate ("Sniff, sniff, ah, here's where I went before.")

I also wanted them to play with the cat regularly in problem areas so she would begin to associate those locations with play instead of toileting.

Some cats outside seemed to be contributing to the problem, for Chantilly became aroused when the cats were prowling around the backyard. She also seemed to be easily aroused by novel smells in the house. Mr. Gardner noticed that when he brought his briefcase home from the office, Chantilly could become quite agitated sniffing around it. Because of this frequent arousal (the outdoor cats couldn't be kept away) we gave Chantilly a short-term dosage of some tranquilizing medication to calm her.

With a methodical program of redirecting Chantilly's attention back to her box, she gradually made fewer and fewer mistakes in inappropriate locations. Over the next two weeks, she began to act as though her box wasn't such a bad place after all. The true measure of success came during week four, when she began to let China use it, too.

So if you see your cat heading toward the litter with a distinct mission in mind, what could it hurt to say "box" at the same time and start building up that association? Any other word or phrase you choose, used consistently at the appropriate time with the same intonation each time, will have the same effect. You never know when this training may come in handy.

Another trouble spot in which training a cat might prove useful is the kitchen counter (or the curio cabinet or wherever the cat insists on leaping). Most people with cats face "jumping up" problems at one time or another. You know the scenario: You've prepared an elaborate meal, set the dining table with the finest china and wine glasses, and created a lovely centerpiece with fresh flowers. Just as your guests seat themselves, Buttons takes a flying leap and lands beside the boss's butter dish.

Many cats are curious; they explore a lot, and they like a nice high vantage point. If you can provide them with an alternate selection of shelves or window seats, they may be willing to give up the kitchen counters and dining room tables.

Of course, it's usually not that easy. Often a cat will be a table or counter jumper for a specific reason, usually spelled f-o-o-d. I remember one kitten, Barney, who was constantly jumping up on the kitchen counters, eating everything in sight, and knocking things over.

I've found that many people's solution to a problem like this is automatically to yell at the cat, which can have the effect of making the cat afraid of the yeller (so the next time, the cat sneaks onto the counters when the owner isn't around). Barney's owners, two young flight attendants, were no exception when it came to yelling. Even as we conferred in the living room of their apartment one Sunday, Barney was up on the counter helping himself to a few leftover lunch crumbs.

"*Scat!*" Julie Fink shouted, and the kitten leaped down, knocking over the salt shaker and disappeared around the corner into the back hallway.

"Well, that certainly got Barney away from the kitchen counter," I said to Julie, "but is he still fond of you when you get near him? Does Barney show much affection to either of you when things are calm around here?"

"No," Kate Norcross spoke up. "We spend a lot of time yelling at him to get him off the table and counters. . . . I guess he sort of avoids us."

"So, let's let the mousetraps be the villains for a change," I suggested, "and you can start being the good guys. Bring the waxed paper over here, too."

I showed my clients how to set some sheets of waxed paper on the counter, with the ends hanging down so that the cats could see it from the ground. Then I set three small mousetraps I'd asked my clients to buy and turned them upside down on top of the waxed paper. The mousetraps were safe, commercially available cat trainers attached to a wide plastic base for easy activation.

Now we had a booby trap that would startle but not hurt the cat. "When Barney leaves his launch pad and lands on the counter,"

I pointed out, "the mousetrap will go off and momentarily startle the heck out of him. And from Barney's perspective it will have nothing to do with you."

"What's the waxed paper for?" Kate asked.

I told her that it served as a signal to the cat on the ground that the booby trap was in place. ("When that's on the counter I won't like it if I jump up there.")

"Because it's the waxed paper the cat learns to avoid, you'll be able to get rid of the mousetraps very quickly," I said. "Let the waxed paper stand for them, and it'll serve the same purpose. Then you can take the paper away for a day and put it back the next day, take it away two days and put it back the next day, and so on, and Barney will gradually establish a pattern of not jumping up there. At that point and from then on, you won't need the paper anymore at all."

After a few trials, this method did the trick as far as Barney was concerned. Within three weeks of my visit, Barney was no longer jumping up on the counter, and after another week Kate and Julie reported that they were able to dispense with the waxed paper.

This method can work for many other cats faced with the same choice—to jump or not to jump. You can use a towel or anything else the cat is not familiar with as a substitute for the waxed paper. It also works well in training a cat not to leave little pawprints all across your newly waxed car; the principle is the same. The only caveat I will offer is that I wouldn't try these scare tactics with very sensitive or nervous cats because it could traumatize them. As always, each cat has to be treated individually, and that's why I go to people's homes, meet the cat, and take a behavioral history before settling on a course of action.

Another common problem young cats present their owners is wrecking the carpet and furniture with their claws. Among the methods of marking a territory is scratching, usually performed on vertical objects such as chair legs, sofa arms, or almost anything made of rattan. Occasionally, an Oriental rug is preferred, but only if it is priceless and/or irreplaceable.

Why do cats do this? For one thing, they have sebaceous glands in their feet, which leave an odor where the cat has scratched. But scratching serves several functions in addition to marking: sharpening the anterior quadradactal appendages (claws); stretching upward or outward with the front legs, as when your cat tugs on your wool rug; and soliciting play (getting owners to play tag, especially when they've finally relaxed in the evening with the newspaper). Scratching increases when a new piece of furniture is introduced into a cat's living area or when the present furniture is rearranged, and it occurs more frequently when other cats are present.

Scratching is usually considered a territorial sign for a conspecific (another cat) or an indication that "herein, I spend a preponderance of my time." Scratching serves as a nice visual signal (during its performance and afterward in the form of scratch marks) and as an olfactory signal. Neither scratching, scratch marks, nor scents, however, seem to inhibit cats' activity in an area that has been marked by another cat.

Fore-claw sheaths are frequently found beneath or even in a marked location (tree bark, for instance), indicating that it was time to trim the nails. However, the rear claws are rarely used in the same manner (compared to a dog's scratching with its rear legs), and declawed cats continue to scratch, although they fail to leave the usual scratch marks. Fortunately, scratching can often be displaced onto preferred objects—preferred by the owner. But of course, without Kitty's acquiescence, displacement will not occur.

If a kitten is just beginning to exhibit scratching behavior, you can sometimes prevent a bad habit from forming by simply interrupting: Say "no!" and clap your hands or just call the kitten to you. This requires some vigilance, and if you can't police the activity, or if the habit is already well established, this method is not as likely to work. Then, aside from declawing (which stops the damage but not the act of scratching), redirecting the action onto an acceptable object is probably the best way to save the sofa.

As with other forms of behavior felines are genetically prepared to exhibit, such as urinating in a loose substance and covering feces, it is easier to displace scratching than to eliminate it altogether. The idea is to make the target (chair leg, sofa arm, and so on) less desirable, and the new, preferred area more desirable to the cat. This is the same strategy one uses to return a cat to the litter box.

The following suggestions are ways to make the target area less desirable:

1. Cover the target area with something that feels different from the material you want Kitty to stop scratching. Examples include plastic carpet runner for the rug or couch cushions (feet-side-up is more aversive than smooth side up, but you can try smooth side up first); for vertical objects try double-sided sticky tape, aluminum foil, or another nonpreferred surface. Many cats prefer soft carpeting for scratching and urinating and reject hard or smooth surfaces.

2. For locations infrequently visited by people, wipe the scratched surface with something that smells or feels unattractive to Kitty, such as chlorine-based bathroom cleaner, mothballs in a box or something else Kitty can't lick, or petroleum jelly.

3. For horizontal surfaces, consider blocking access to the location or (more aversive) placing on the scratched material the cat-oriented, easily set mousetrap designed to keep cats off such surfaces. These suggestions can be used as alternatives to 2.

4. As a last resort, use a commercially available electrified plastic carpet runner or a small security device that produces a loud noise when Kitty walks close to the scratched surface and disrupts a laser beam. This is likely to keep some cats from even entering the room in which the scratched surface is located.

But I stress that these are last-resort strategies because while effective, they can be upsetting to sensitive felines. You know your cat's temperament.

In addition to making prohibited areas unattractive, try to get the cat interested in alternative scratching areas. Ways to make the new preferred object more desirable include:

1. Place the new object (such as a scratching post) next to the inappropriate object. This works well if Kitty persists in going to a specific location in the house to scratch, rather than wherever the rattan is located.

2. Select a scratching post made from material for which Kitty has shown a preference. Many cats prefer to sink their claws into tree trunks, complete with bark (or smaller versions, as in wood logs). The "trunk" should be long enough to allow Kitty to fully stretch up vertically; effective materials include corrugated cardboard or a "paper roped" chair. Some cats are lucky enough to have their own chairs complete with pine or rattan legs and tightly wrapped chair backs and seats made from paper rope. The chair is effective in displacing both vertical and horizontal scratching. Other materials include vertically or horizontally ribbed carpet that can be wrapped around and stapled to a scratching post and replaced as necessary.

3. Place at the top of the scratching post a Ping-Pong ball on a string, an anchored stuffed mouse, a catnip treat, or other toy. The idea is to encourage stretching and other activity that will direct Kitty's claws toward the scratching post. Play with the cat at the post.

4. For cats who like to scratch a specific material, no matter where it is located, gradually move the new object to a desirable location (desirable from *your* point of view—but be careful not to insult Kitty by placing it where no one can see it; remember,

scratching is often used as a visual marker.) Moving the scratching post short distances of a foot or two every five to seven days usually works well.

You might also create a private space for your cat, safe from grabby toddlers and dogs who like to get up close and personal. You can buy a commercial cat door to install for going outdoors. Inside, designate a single room—whether it's a pantry, mudroom, or unused storage area—for your cat's privacy. Install a door with glass panes and remove one of the bottom panes so that the cat can hop inside and still see what's going on back in civilization. Soon she'll be leaping back and forth and loving her new private space.

What if your cat doesn't scratch, jump up on the table, or ignore the litter box? Let's say fighting with fellow cats is the activity of choice. What's an owner to do? One of my clients had inadvertently trained his cat to respond to a signal that would at least temporarily stop the aggressive cat in his tracks while we worked on the cats' relationship with each other. Perhaps your cat will respond to some kind of special signal, too.

Rescued from an abusive family where he had spent much of the time outdoors, the Maine Coon was renamed Napoleon and integrated into a platoon of three other males: Patton, Caesar, and MacArthur, all indoor cats. He quickly found himself under the gun again. This time it was his own species that was out to get him, and he had nowhere to run.

Caesar—the oldest of the cat quartet at seven years—was making two-year-old Napoleon's life miserable, ambushing and attacking him relentlessly. So I sat down with the owners, Sgt. Bill Warren and his wife, Molly, to put together a plan to win the peace.

After hearing the details of several skirmishes and taking a tour of the battlefield, I was able to get a clear picture of what was going on—and where some effective changes could be made.

From what the Warrens had told me, it seemed that Caesar

was trying to maintain his leadership role in the group and hadn't exactly rolled out the welcome mat for the new cat. But with so many cats, it would be easier to deal with changing some strategic locations than with relationships.

"Trying to get Caesar and Napoleon to be friends might just not be practical," I suggested.

"Just so they can peacefully coexist," Molly said. "Where do we start?"

"How about with your 'weapons,'" I said, half jokingly. "Were you able to find some of those dime-store clickers they used to have when we were kids?"

"Yep." Bill fished the red metal clicker out of his pocket. "Of course, the price is no longer a dime."

The clicker was to be the final evolution of a signaling device that had started with a wet towel. Bill had told me in the initial phone call that when he came out of the shower one day and found the cats brawling in the bedroom, he had quickly taken the closest thing at hand—the towel wrapped around his waist—and snapped it smartly at the felines. This really startled Caesar, and he ran out of the room, while Napoleon initially froze, then retreated under a desk.

Later, Bill had discovered that a snap of his fingers approximated the sound of the snapping towel and was proving effective as a signal for Caesar to curtail hostilities. He'd just stop whatever he was doing and look at Bill.

I wanted to take advantage of this already existing behavior modifier and asked the Warrens if they'd be willing to use this signal quite a bit during the training program. I had urged Bill to find a mechanical clicker and thereby save his fingers. Molly was also armed with a clicker.

We talked about the many adjustments Napoleon had been asked to make in the past month or two, and the Warrens agreed that the new cat should be permitted the comfort of having the bed to himself instead of always having to retreat under the desk or bed whenever Caesar threatened him.

This meant the clicker would have to be used to keep Caesar from entering the bedroom, especially if Napoleon was there. If there was already an attack in progress, the clicker would be used to interrupt it. Eventually the sound of the clicking would cause Napoleon to run to his safe area—the bedroom—and Caesar to run away from the bedroom in reaction to the commotion. This approach, I was to learn in follow-up calls to Bill and Molly, actually worked quite well and became a ritualized pattern for the two cats.

In order to lower the number of cat altercations, I asked the couple to show me where any other attacks on Napoleon had taken place. One major problem, they said, was the litter-box area—where they had seen Caesar lying in wait to ambush Napoleon when nature called. The box was in a little laundry room at the bottom of the basement stairs, and a cat couldn't see if another feline was there until it was too late.

Given this treacherous condition, Napoleon seemed to have worked out a strategy for avoiding attacks by Caesar at the litter box. His coping tactic was to wait for MacArthur to go downstairs, then follow quickly behind him to avoid being ambushed. He absolutely wouldn't go down there without the other cat running interference for him. While this was an admirable way of trying to cope, it could lead, I felt, to a litter-box problem. Napoleon just might begin to avoid the stressful location altogether and go elsewhere, not necessarily in a box.

So Bill and Molly and I were able to find a better spot for the box in the basement, where the cats could see each other coming and escape if necessary. I asked them to add another box upstairs in the back hall to lessen the chances for confrontation.

"Where else do a lot of attacks occur?" I asked Bill.

"Around the food bowls," he replied. "Caesar makes Napoleon

wait until last for his food, and if Napoleon surprises him at the food dish, Caesar runs him off. We feed them right here, around the corner from the breakfast nook," he pointed out.

I had to smile. As with the litter box, the food bowls were in the best possible location—for an ambush.

"See anything risky about this location?" I asked the sergeant.

"Yep. I guess we need to remove that surprise element from the eating routines, too," Bill responded, picking up the bowls.

"I see what you mean," Molly added. "Every time Napoleon turns the corner when Caesar is eating, Caesar runs him off. Napoleon would probably make himself scarce if he could see that Caesar was waiting to pounce on him."

"How's this?" said Bill, setting down the food and water dishes in a corner of the kitchen. "Now the cats can see who's here eating all the way from the living room."

"Terrific!" I told him. "That should do it—let me hear from you next week."

Before ending the session, I asked my clients to consider one more important element in their battle plan.

"We've talked a lot about 'poor Napoleon' today," I pointed out. "But let's not lose sight of the problems Caesar has been facing trying to remain on top."

Molly and the sergeant nodded. They were concerned about how their old friend would take to so many changes.

"I'd like Caesar to know that he is still a good cat, so as soon as he stops an attack when you say 'no' or use the clicker, make sure you praise him."

"Won't this bad-cat/good-cat routine confuse him?"

"It shouldn't. One of the most basic principles of conditioning behavior is that animals respond to the praise of the most recent behavior they engaged in. So what you're praising is the cat's stopping the misbehavior, and they seem to understand that."

With positive reinforcement added to the mix, the hostilities in the Warren household soon came to a halt. I was especially impressed with the progress Napoleon was able to make, considering his background as an abused cat.

It takes some effort to replace frustration and punishment with teaching desired behavior and to offer consistent praise to a cat for a good job, for improvement in its behavior, or just for trying. But it usually pays off handsomely.

6

Beware of
the Attack Cat!

In all the years I've worked with pet problems, I never met an owner more terrified of her own cat than Gail Barnes was. And the minute I opened the door to her Atlanta townhouse, I could see why.

Usually, cats do one of two things when I appear for the first time. If they are very confident and social, they cautiously approach to investigate. If they're more shy or solitary, they invariably use an escape route when the doorbell rings, and if I'm lucky I may get a short exploratory visit or at least a glimpse of a nose or paw from under the sofa before the end of the consultation. Occasionally there will be a crazed one like our friend Whiskers, who was simply too busy tearing around to notice the doorbell or a stranger.

But Midnight was different. As I entered the foyer, I took one look at him and braced myself. The three-year-old tabby was crouched tensely on the staircase banister, ready to spring. He leaned forward on his haunches, glaring at me with wide, dilated eyes; all his body language indicated that he might lunge at any moment. This was definitely not normal cat-to-stranger behavior.

"Come on in," said Gail, with a nervous glance over her shoulder at her pet. Suddenly the cat leaped off the banister and raced down the staircase to the lower level, and Gail quickly shut the door behind him. I found myself breathing a little easier.

We sat down in the living room, and I began to take Midnight's history. It was pretty uncomplicated. Gail had acquired her pet when he was three months old, and the cat had done nothing remarkable in his life until four days earlier, when he had suddenly, "for no reason," viciously attacked his owner, biting her on the leg and scaring her half to death.

A second attack from which Gail had barely escaped unscathed had taken place two days earlier, and that's when Gail had called her vet pleading for help. The young woman had no idea why the cat had suddenly turned on her, and she was upset and frightened.

Gail rolled up her slacks to show me the healing claw wound. People usually want to show me the damage to their person or property to make sure that I see that they haven't blown it out of proportion. When I go to a house where the cat is making mistakes, I know I'm going to be shown the telltale spot on the rug. They'll exclaim, "Have you ever seen anything like that?" And since this is my profession, I probably have. But they are usually relieved to hear that they are not the only ones with the worst-behaved cat in the world.

"I'm sorry you and Midnight have had this terrible experience," I told Gail. "I'm pretty certain I can help you, but it's going to take a bit of detective work. Let's go through the events leading up to the first attack and see if we can figure out what's making Midnight so aggressive."

Gail nodded and took me over to the sliding glass doors leading to her small patch of yard outside.

"Well, that first afternoon, I had just cleaned Midnight's litter box, out there," she said, pointing to the terrace. "I put in the new litter, brought the pan inside to the kitchen, and right away Midnight jumped up on top of the lid and started hissing and growling at me.

And then he just flew off the box and started biting my foot and ankle like a wild animal."

"What did you do?" I asked.

"I screamed, and it must have startled him because he let go for a second, and I ran to the bathroom and locked myself in," Gail said. "I took care of the wound—my leg was bleeding pretty badly— and I stayed in there a little while longer, about a half hour altogether, and then I snuck out and ran to the bedroom. Midnight ran after me and sat outside that bedroom door growling for at least an hour."

"That must have really shocked you," I broke in.

"It did," Gail admitted. "I felt like I was in a horror movie."

"Well, if it's any consolation," I said, "Midnight's persistence isn't unusual in this kind of aggression. Most people aren't aware that cats stay aroused and irritable for a long time after an aggressive bout—sometimes up to two hours or more—whereas dogs calm down after fifteen or twenty minutes. The same is true with cats that have been frightened. I once saw a cat that wouldn't move from a treetop for four days, until it became so weak that it finally fell down. Maybe that's where the term 'scaredy cat' comes from."

Gail smiled miserably. My feeble attempt at a joke hadn't helped much, and the young woman definitely needed to lighten up. She was going to have to be able to live and work with the cat, but right now she was still a bundle of nerves. Not that I blamed her.

"You did the right thing to stay in a safe place that long," I assured her. "What happened next?"

"Well, I actually waited *another* hour, and then I came out and he seemed to be OK. He *was* OK until two days later."

I asked her to tell me about that incident.

"OK," she continued. "I'd just finished doing some gardening out back—potting some geraniums on the terrace—and I remember Midnight was watching me through the window, just sitting there acting perfectly normal. But when I came in to wash my hands,

Midnight jumped up on the countertop right next to me and start-ed that hissing again."

Gail got up and moved to the kitchen, and I trailed after her. "I was standing here, next to the refrigerator, and he jumped down to the floor, hissing and growling and glaring at me. I just backed along to the door, like this," she demonstrated as she edged along to the slid-ing doors in the living room, "and opened it quickly and went out and slid it shut. At the same time, Midnight lunged at me but banged into the glass. I stayed outside for a while until he seemed to calm down, then snuck in the front door. "He hasn't acted really like himself since then. You saw him when you came in." She gave a discouraged shrug of her shoulders. "So, what am I going to do?"

"Well, let's get to the bottom of this first," I said, "then I'll give you some suggestions. Now, both attacks took place immediately after you came in via the sliding glass doors to the backyard, right?"

"Yes, I guess they did."

"Does Midnight ever go outside?"

"Oh, yes, for a little while each day."

"Are there other cats around your yard?"

"Well, not normally . . . but lately there has been one—I don't know where he comes from—he's a big orange tomcat." She chuck-led. "Midnight chased him away the first time he showed up."

"Is it possible that Midnight ever tangled with this cat?"

"Well, I suppose so. . . . He did come in looking very scruffy, eight or ten days ago."

"OK. When you were outside cleaning the litter box and plant-ing geraniums, did you use a watering can or the hose, so that the ground got wet?"

"Yes, both times I used the hose."

"All right, Gail, I think I can make an educated guess now about what caused Midnight's strange behavior."

I closed my notebook and offered my hypothesis. "You very likely brought the odor of that tomcat in on your shoes after working outdoors. Wetting the ground with the hose mixed his urine with the soil and helped it rise to the surface."

I explained that the chemosignal or odor in the urine—unnoticeable to Gail—was probably what drove Midnight up the wall. He associated Gail with the odor and released his attack onto her. (We behaviorists often refer to stimuli as having a "releasing" effect on motivated behavior such as aggression.)

"He couldn't attack the orange cat," I went on, as Gail listened intently, "so—unfortunately—Midnight attacked you instead. This 'rage' reaction to the urine allowed him to release his aggression on the closest moving thing. It can be very dangerous and vicious, as you've found out." She nodded.

Fortunately, rage is uncommon in cats, yet you'd think cats would be able to inhibit aggression toward a caretaker. But they don't always, so we know that whatever this chemosignal is, it's a very powerful stimulus to aggression.

"So, are you ready to help me solve this problem?" I asked my client.

"I'll do anything I can."

I asked Gail to try to determine who the orange cat belonged to and in the meantime to be sure to wash her hands and take off her shoes outside before she came in. She also agreed to have the carpeting in her townhouse cleaned in order to eliminate any trace of the odor that was still disturbing Midnight.

According to Gail, she and Midnight had always enjoyed playing rather roughly, or playfighting; now I suggested that she substitute gentler playing—petting, throwing toys—until the problem was well in hand. There was no sense in further arousing the cat, and to that end we also agreed that Gail should keep Midnight inside until the problem was resolved. I asked Gail to call me with a progress report in a week.

Gail reported the following week that she had asked all the neighbors about the orange cat, and no one knew who owned it.

The general opinion seemed to be that it was a stray. Gail noticed it still outside and worried that if Midnight saw it, he might react aggressively and hiss as he did before. Finally, Gail had called the local animal control officer, and soon the tomcat no longer appeared.

Having done everything she could do, Gail played gently with Midnight and watched him carefully. Two weeks later—to the immense relief of his owner—Midnight was back to being the good-natured pet Gail knew and loved. By the time Gail's bites and scratches healed, she was confident that there would be no more instances of irritability or rage in Midnight.

"Cat bites and scratches?" you might think. "Big deal!" Compared with the fifteen to twenty fatalities caused each year in America by dogs attacking humans (usually children), cat bites and scratches are not taken very seriously by most people. You've probably seen this sign in a pet store or novelty shop:

THIS HOUSE IS PROTECTED BY AN ATTACK CAT.

The thought of sweet Snowball or timid Tiny deterring prospective burglars is droll. But at least half a million Americans each year, and many, many more whose experiences go unreported to health agencies, aren't amused. These people—including burglars, I suppose—have been bitten or scratched by their own cats or others, most of them seriously enough to require first aid or medical attention.

And the story doesn't end there. Public health officials estimate that as many as 50 percent of cat bites and scratches become infected, a rate much higher than for dog bites. The national Centers for Disease Control and Prevention and other facilities similarly report a higher incidence of rabies in cats. Currently making a comeback on the East Coast, the disease is transmitted by the saliva that accompanies the bite of an infected animal and is always fatal to unvaccinated cats. With cats becoming more popular all the time and being far less likely than dogs to receive rabies shots, these problems will be on the rise as well.

So, all things considered, I'm likely to take cat biting and scratching pretty seriously, and I try to help owners eliminate it as quickly and thoroughly as possible.

Aggression is the second most common cat behavior problem I encounter, with the litter-box problems I outlined in Chapter 2 accounting for the majority of my calls. Sometimes common sense or a telephone conference with an animal behaviorist or a knowledgeable veterinarian can be all that's needed to turn around certain simple cat misbehavior problems. These can be tackled by the pet owner without fear of danger or regression. *But aggression isn't one of those problems.*

Diagnosing and treating aggression is often tricky. Biting and scratching behavior can be the result of fear, playfighting, redirected or displaced aggression, predatory behavior, or a petting-and-biting syndrome, or a mixture thereof. Physical punishment, the incorrect use of antianxiety drugs, or home-remedy attempts at desensitization and counterconditioning programs are likely to backfire and reinforce the aggressive behavior, making matters even worse.

So it's best to leave these procedures to a certified applied animal behaviorist, who will take a careful behavioral history, consult with the veterinarian on any medication necessary, and be able to identify and have the owner remove arousing and "eliciting" stimuli. Then the owner will be shown how to help the cat replace hostile behavior with more socially acceptable and less dangerous activities.

While cat aggression problems always need to be diagnosed and treated by a professional, cat owners can and should be aware of the danger signals a cat may display when aggressive or fearful. Cats look and act differently depending on whether they are in an offensive or defensive mode, but cues for either mode can signal coming aggression.

Offensive postures include staring with constricted pupils at the potential victim (cat or human) with neck and whiskers thrust forward and ears rotated backwards. The cat's hair may stand on end, particularly along the middle of the back. The tail is likely to be held slightly away from the body and bending downward, or the tail may be vertical and the rear end elevated. The cat faces its victim straight on.

When a cat is in a defensive posture, usually facing an enemy sideways, he will try to make himself as small and insignificant as possible. Therefore, he crouches down, pulls in his head with ears flattened, and wraps his tail around in front of his body. In spite of himself, the defensive cat may find all his hair standing on end, an indication of physiological arousal.

The defensive cat may roll over on his back, frightened but ready to defend himself (except in the case of a female in heat; then she will roll back onto her feet). This is not the same as being submissive; the cat is actually preparing to attack. A sideways stance is also defensive. If the aggressor is facing the victim and staring, the attacked cat will avoid eye contact and/or pull his head in.

With defensive postures, dilated pupils are common, triggered by the autonomic nervous system, which regulates those bodily functions that aren't consciously controlled, such as heart rate and smooth muscle activities.

The cat's tail position is a good barometer of what he has in mind. A vertical tail position is like a greeting, or for a female in heat it is a sexual approach. The tail held high also precedes play. If the tail is concaved and curved away from the body, it signals defensive aggression. If it is held low, it is also likely to precede aggression. And like a dog, a cat with his tail between his legs is showing exaggerated submission, or fear. *A fearful cat is just as likely to end up attacking as an aggressive one is.*

Unfortunately, in real life cats don't always exhibit these pure textbook behaviors. Many cats will display several elements of both offensive and defensive aggression during any one incident. These mixed messages are likely to puzzle even the most observant cat lover. Then there are the cats who show no warning signs at all and whose actions are next to impossible to predict.

Regardless of the cat's signals or the cause of the attack, there are some general rules relating to aggression that can be helpful to the owner in heading off this type of behavior.

First of all, an attack toward a person could be predatory, or playing, or conspecific (in which the person is treated like another cat). But the type of aggression most often encountered by owners will be playfighting—especially with cats up to a year old—or redirected aggression, in which a cat (like Midnight) attacks because something else has upset him but isn't available to attack.

To prevent playful aggression, you should not handle your kitten roughly or encourage her to wrestle or scratch or bite your hands or feet. Set her down if she does so. Handle and pet the kitten calmly after she tires of playing. Physical punishment for scratches or bites will encourage more playfighting or make the cat fearfully aggressive.

The best thing an owner can do when aggression is frequent, or aggressive postures are evident, is to leave the cat alone (especially if one is in danger of being viciously attacked, as in redirected aggression). Then call for professional assistance.

If cats are fighting each other, playful aggression usually has nothing to do with it. Fear or territorial imperatives are usually at work in these instances. Always separate fighting cats by giving one an escape route, if possible. But for your own safety, don't reach for the combatants until after they are no longer aroused; you may not want to handle them for an hour or two. They should remain apart until the problem can be addressed by a qualified behaviorist. Don't assume that the cats will work it out on their own. After a serious fight, odds are they won't.

Although both sexes can show aggression, intact males may specifically stalk or ambush other males in addition to defending their territory. Neutering will be beneficial for most such intermale aggression problems, but a behavior treatment program should be followed simultaneously.

If your cat growls and hisses at the neighborhood cats outside the window, as Midnight did, or chases the new feline member of the family all over the house, chances are it is displaying territorial aggression. While a cat will not typically react aggressively to a strange person entering the family's yard or house, you can almost always expect some territorial aggression with feline intruders. And, unfortunately, not just with outdoor cats it has never encountered before. Another cat in its own house ("my turf") can be equal cause for alarm to a territorial feline.

In one case I handled, the resident cat was treating the new kitten as though it were a threatening stranger. Before our appointment, I had asked the owners, two young women, to gather a number of items that might be useful in working with the cats. Through trial and error, behaviorists have evolved a list of readily available household products that serve as tools for dealing with pet behavior problems. When I arrived at my clients' apartment, they let me know that they found this laundry list most puzzling.

"One apple-orchard–scented room deodorant."

"Check."

"A set of car keys."

"Check."

"One container of fish-flavored cat treats."

"Check."

"A piece of aluminum foil."

"Check."

"One fringed throw rug."

"Check."

"That's it," said the blond young woman to her red-haired

friend. She turned expectantly to me. "There you go, Dr. Wright; I guess that's everything."

Jennifer Dixon and Kristy Salinger, Atlanta roommates in their early twenties, exchanged curious smiles as they put the last of the items onto the pile on the living room floor. They were obviously having second thoughts about consulting me to deal with their two cats' difficulties; their landlord's advice, "Smack 'em with a newspaper," probably sounded pretty logical up to this point.

"You may be wondering how this pile of stuff is going to help solve your pets' problems," I ventured with a grin.

"Uh, yeah," Jennifer muttered skeptically, as Kristy nodded.

Their two alley cats, Alpha and Blue, needed all the help they could get. Alpha, a feisty thirteen-year-old, had taken exception to the entrance of eight-month-old Blue into his domain. In addition to attacking Blue at every opportunity, the senior cat was spraying indoors and going nuts when his owners left for the hospital where they were nursing students. Even as we conferred, growling and hissing from the kitchen punctuated the conversation.

"Shall we begin on the landing?" I suggested. We should. "Please bring the throw rug, the keys, the cat treats, the foil, and the room deodorizer."

The landing, five steps up the staircase of their sparsely decorated apartment, was the stage for much of the undesired drama in my clients' lives. From Alpha's perspective, they had told me, it was where the action was. It was the favorite attack-the-kitten zone, the favorite growling zone, the favorite get-upset zone, spraying zone, and go-crazy-when-the-owners-leave zone.

I laid the throw rug on the landing and opened the deodorizer borrowed from the bathroom. Soon the smell of cinnamon spices filled the stairwell.

"The idea here is to make this area feel and smell differently to Alpha," I explained to the young women. "The new texture of the cotton throw rug instead of the wood flooring should help inhibit the spraying in this location. You've done a good job of keeping the area clean, but the deodorizer should change any odor cues he's picking up here and give the area a whole new ambiance. Although it's the same location, the misbehavior shouldn't automatically kick in here because now it's a 'new' spot."

"OK," said Kristy, obviously unconvinced, as Alpha mounted the stairs to investigate what was going on.

"Now, you said Alpha responds to the sound of the keys?" I handed them to Jennifer.

"Yeah, he usually stops attacking Blue if I jingle them near him," Jennifer confirmed.

"All right, so let's keep a few sets of keys around the house to head off these confrontations," I advised. "This will work best if you combine it with some play behavior. Do you play with the cats?" I suspected not; at least there were no toys evident around the apartment.

Kristy and Jennifer exchanged guilty looks. "Not really," Jen answered. "We don't have much free time."

"OK," I went on, "well you can take a moment when it looks as if Alpha is preparing for an attack. Jingle the keys first to get Alpha's attention, then throw a toy across his visual field from left to right to elicit a chasing response." I crumbled up a bit of tinfoil and demonstrated; the black cat went bounding after it.

"If he uses up some of that energy in playing, he'll have less left to attack the kitten," I remarked, as the young women nodded.

Although the motivation for territorial aggression is different from that for playful aggression, the same kinds of behavior are exhibited: chase-bite-run combinations. Thus the idea was to release the attack onto a play object rather than onto Blue.

Next we went over the use of another distractor, the cat treats my clients were to use when they were leaving each morning. By throwing the treat away from the door when they went out, I explained, they would take the cat's attention away from their leaving and interrupt the behavior pattern that led to Alpha's attacking Blue.

"In other words, change Alpha's focus on your going away, which usually gets him hissing and racing around and running into Blue, to one of 'wow, there's something to eat over here; let me concentrate on that.'"

Because Blue didn't compete with Alpha for food (subordinate cats rarely do), Alpha was able to have his cake and eat it, too.

"One more thing about Alpha," Kristy said. "You can't look him in the eye or he starts to growl."

"Then try not to look him in the eye," I smiled. "That's too much assertiveness for him to handle right now."

Soon the situation eased, and as the two women held treats at their eye level before tossing them to Alpha throughout the day, Alpha gradually became more accepting of eye contact.

This case was typical for cats that have not been properly introduced (the kitten had just been brought home one day and plopped down beside the older cat) and have gotten off on the wrong paw. A gradual introduction, with the cats separated by a door or window for the first few days, might have set an easier course for those two fighting felines.

Generally speaking, each instance of territorial aggression is different, due to the differing age, sex, living conditions, and previous experience of the felines involved. Not all cases of territorial aggression are resolved so easily; and in some cases, unfortunately, placing one cat up for adoption may be the only real solution.

Aggression that takes the form of predation looks different from aggression toward another cat and usually spares humans as well. When stalking a victim, experienced cats will slink along with their tails swishing back and forth, bringing all their senses to bear on tracking the prey. Sometimes predation is not even classified as aggression, because it involves a different motivation and activity in a different part of the brain. You've probably heard that it is the cat's "instinct" to kill birds and mice. But applied animal behaviorists today don't really talk about instincts, as typical feline behavior may have a number of sources of influence, and the word "instinct" has been overused to such an extent that it's not very useful.

What will the typical feline attack, and how? That depends mostly on the size of the prey. Just about anything smaller than the cat can be fair game. A cat is more likely to attack moving things, rather than still ones, and strange creatures rather than those with which it has been reared. Kittens who have been raised with a mouse generally won't attack the mouse. In fact, there have been cases of mice and rats actually reared by cats.

A cat will attack a frog by leaping onto it and then possibly eating it. With a mouse, they will slink close to the ground, dart toward an area where there is something to hide behind, make another dart, and on the third dart, chase and attack. The cat is likely to take its paw and hit the mouse so that it tumbles over, and then bite it at the nape of the neck or throat.

Often a cat will continue to toss a dead mouse or rat in the air long after it is dead. Owners are usually put off by this seemingly cruel and sadistic behavior. But the larger the prey, the more fearful cats tend to be, and the more likely they are to play with the prey, even when dead.

Cats will hunt, attack, and kill prey regardless of whether they are hungry. (Whether the cat will then eat the prey depends on how it tastes.) If you think you can keep a cat from attacking birds by feeding the cat a lot, think again. Cats will start hunting immediately after consuming a meal. They will attack nearby mice while already eating another mouse.

When kittens are four or five weeks old, their mother will bring them a not-quite-dead mouse (if she has access to rodents). They will practice pawing, biting, and killing it. Then she will show them how to attack and kill one. The kittens will practice the predatory attack during play so that they get better at what they're doing. Motor-action movements are repeated and become fine-tuned. These kittens usually end up much better at killing prey than cats raised in a pet shop or with a breeder, where they and their mothers have not been given the opportunity to hunt.

The practicing of predation coincides with playfighting among the kittens. Littermates, the mother, or other adult cats are targets of a pounce, belly-up or stand-up, to initiate play. After some rough-and-tumble action, play sessions usually end abruptly with a chase or vertical leap. Behaviorists have observed that single kittens play more with their mothers than play alone or with objects, and this play involves considerable tail batting and stalking.

At about four months of age, social play becomes more aggressive—and this can lead to trouble when an owner doesn't understand why his loving pet is attacking him every time he walks down the basement stairs. Sometimes the behavior a playfighting cat directs toward her owner becomes so ritualized that it is pretty easy for an animal behaviorist to recognize. Such was the case with Nacho, a gray striped female who relentlessly—and predictably—attacked her owner, Martin Jessup, every night as the man went into the master bathroom to prepare for bed.

"Nacho hides under our dresser and then jumps out and attacks me every time I walk by," Mr. Jessup complained as we spoke on the telephone. "It's really getting on my nerves. And," he added ruefully, "my feet are getting pretty scratched up, too."

When I met with Mr. Jessup, I explained that the ritual needed to be interrupted. Putting something under or next to the dresser so Nacho couldn't leap out and start her biting ritual was the logical first step. And I had another idea.

"What is Nacho's favorite treat?" I asked Mr. Jessup.

"That's easy," he replied without hesitation. "She loves yogurt."

"OK. Let's try putting a dish of yogurt out for her as a bedtime snack, just about the time you're ready to retire for the night. Set it near the bedroom door, then you go the other direction, into the bathroom." This activity served two purposes: It kept Nacho away from the attack zone, and it established a competing motivation— the yogurt enabled Nacho to recognize Mr. Jessup as a source of food rather than a moving target.

The next week, Mr. Jessup told me that the yogurt had definitely distracted the cat at the crucial time. But when Nacho wasn't eating, she was still under the dresser waiting patiently for her victim to walk by, a red flag for a resumption of the activity. Mr. Jessup hadn't blocked off the dresser because it seemed too much trouble. And his wife didn't want a bunch of stuff cluttering up the space under her furniture. I (rather irritably) replied that the couple was not likely to see a lot of improvement until they had come to grips with this part of the problem.

A week or two later Mr. Jessup checked back with me.

"The problem's solved, Dr. Wright!" he exulted. It seems that a previously ordered bedroom set had arrived. And, by sheer luck, the couple had ordered a Mediterranean-style dresser that sat directly on the carpet. There were no legs, therefore, no ambush space and voila—no more ambushes. It was a lucky break, and after my usual three-month follow-up, I was able to close the file on the little outlaw Nacho.

While playfighting is a common form of aggression that owners have little trouble understanding once it is identified, cat lovers are often shocked when their favorite feline suddenly attacks the hand that pets it after seemingly enjoying the first several strokes. Their confusion is understandable; cats are supposed to love being petted.

"No one is really sure why," I told Kim Belkin, whose tabby Inkspot had been biting when she petted him, "but you're not alone—

many cats suddenly turn on the person who's petting them." One theory is that the cat is behaving as though another cat were grooming it—which it will tolerate for a short period but will eventually terminate by lashing out. Animal behaviorist Ben Hart recognized this "petting-and-biting syndrome" when he described "the three-stroke cat." Inkspot, it seemed, was a four-stroke cat. If Kim was petting and stroking the cat, we found, he would accept four strokes, and then suddenly he would nail Kim, biting her hand and clawing her arm.

Luckily, Inky was one of those cats whose body language was fairly dramatic and could be interpreted with some certainty to predict an act of aggression. His eyes would dilate and his whiskers rotate toward his victim just before the attack.

"I think we can improve the petting situation," I told Kim, outlining the program she was to follow: Make sure Inkspot was lying calmly, begin to stroke him, and count the strokes—one . . . two . . . three . . . four. . . . When Kim was about to do the fifth and could see the eyes dilate and the whiskers come forward (which was Inkspot's way of saying "Let's see if I can feel this hand before I bite it") then she was to stop petting. Unfortunately, not all cats rotate their whiskers forward before biting, but for those that do, it is a useful sign to be aware of.

I told Kim to start again when the whiskers became less flared or went back. If after the fourth stroke the whiskers didn't come forward, add a fifth stroke, and then stop. Eventually she would be able to sneak in a sixth, and so on. That way, the cat would gradually get used to more and more stroking over a period of several weeks.

This gradual approach proved successful for Kim, and within six weeks she told me she was able to stroke Inky eight or ten strokes at a time without eliciting a bite.

In these cases, it's almost as if the cat were counting, because clearly some cats will allow that third or fourth or fifth stroke, but on the next one, they will attack every time. And, of course, many cats can be petted for what seems like hours and remain completely docile. The difference among individuals is inexplicable, and one of the many fascinating little quirks of feline behavior.

While the motivation behind the petting syndrome remains something of a mystery, animal behaviorists do know that the great majority of cat bites and scratches are defensive—they happen when someone is reaching for a cat, picking it up, trying to hold it when it wants to be put down, and so forth. When I did a study of hundreds of cat bites in Dallas several years ago, I found that strays (unowned or unclaimed outdoor cats) were the biggest source of cat bites, and women—perhaps trying to pet or feed them at the kitchen door—made up the greatest victim pool.

My research showed that bites most often were inflicted on the hand and fingers, and next most often, on the arm, as the person reached out for the stray. Whether the cats were reacting to potential entrapment, the invasion of their "personal space," or the fear of a cuff on the head is impossible to say. But reaching out appeared to trigger the aggression.

Owned cats can act this way, too, and in the case of Coco, the defensive aggression was aimed primarily at the male half of his owner team: Al Riley, a retired salesman. The four-year-old Himalayan had begun objecting to being picked up and was now attacking Al in the sunroom whenever he put on or took off his shoes, walked across the room, or sat on the couch alone or with his wife. Coco's behavior was characteristic of defensive aggression. He would begin with a swipe of the paw and go on to crouch, hiss, growl, scratch, and bite whenever Al came near him or tried to pick him up. But now the defensive aggression appeared when the owners were merely in the same room, not even trying to pick up Coco anymore.

Though cats that behave like this are frightening to those around them, they generally respond well to a behavior-modification program.

A combination of avoiding the cat in the sunroom at times when the attacks were most frequent, displacing his aggression onto a thrown toy before he had an opportunity to attack, and, to head off aggression, watching for the predicting signals I outlined earlier in this chapter yielded good results within two weeks. Frequent playing with Coco and avoiding confrontations in high-risk situations solved the problem in short order.

I've found more aggression problems in purebred cats than in mixed breeds. Sometimes behaviorally undesirable genetic qualities are passed on in purebreds. The offspring of parents of different breeds tend to be very good cats.

You don't find many untreated aggressive cats that have survived past their first couple of birthdays, because owners typically get rid of them if they're attacking people or other household pets. Often an aggression problem is developmental, linked to the stage that occurs from the ages six months to a year and a half, when the hormones are beginning to flow and sexual maturity is reached. That's when owners usually set out to correct the problem or make the tough decision to give up the cat. And unfortunately, many desperate people end up simply putting the cat outdoors and leaving it to fend for itself.

On the other hand, if the sexual-maturity danger point is passed without aggression problems—and nothing else happens to upset the apple cart—cats are more than likely to become pacifists, settling into rather predictable routines.

7

Top Cat, Bottom Cat: Challenges of the Multiple-Cat Household

Anytime you have more than one cat in a household, relationship problems are likely to materialize. Sometimes, the cats will spend the better part of their waking hours jockeying for position and access to desirable resources—you included. If that doesn't work, they will fight. Or, instead of aggression, there's everybody's favorite: spraying. As always, some cats live in perfect harmony, and some cats cannot tolerate sharing a household with another feline. Those are the ones I am called on to help.

If a cat is particularly susceptible to stress, the introduction of a second cat may be enough to break down his coping systems and lead him to urinate or defecate outside the litter box or, more likely, to spray. Spraying is not an elimination problem but a marking problem, related to territoriality, sexuality, stress, or another less common factor.

Spraying is one of several ways a cat marks his territory. A cat will establish a territory, that is, an area he roams throughout and will be prepared to defend (as opposed to his home range, which extends to the farthest point away from home base the cat will normally go, but not necessarily defend). A cat marks his territory as an indication to other felines that he spends a lot of time in that location, or at least that he has been there in passing.

Marking can take several forms. As I mentioned in Chapter 5, cats will mark by scratching and leaving a scratch on bark or another surface, especially on vertical objects that will be visible to the other cats. They also leave their calling cards by means of some odors from scent glands in the pads of the feet, or they rub up against objects with their body and the scent glands on their head.

Cats also mark by spraying. The cat will back up against any vertical object, lift his tail, and spray urine as the tail quivers. This kind of posture indicates marking rather than elimination motivation. If there is a new cat in the vicinity, the resident cat may be cordial and avoid picking fights, but at the same time spray all over the house to emphasize his presence to the intruder.

What effect does marking have on the cat who is being notified "This is my turf—go away"? Probably very little. The second cat will generally just sniff and walk on—the equivalent of a shrug and a "so what?"

A cat encountering another's spray will not normally make it a point to mark over it in the same spot as a dog does. He may, however, pick a spot nearby on which to leave his own calling card. In any case, the carpet is ruined. It is easy to see how spraying cats can quickly turn a household into a disaster area.

Here are a few facts that you might want to consider if you are tempted to take in a whole litter of kittens or to skip neutering the ones you've already got, with the view "the more the merrier."

Males are more likely to spray than females, especially males that have not been altered. Castration is likely to reduce or stop the spraying.

Researchers have found a direct relationship between the number of cats in the household and the probability of spraying by one or more of them. One survey of 150 cat owners revealed that in a household with one cat, there was a 25 percent chance it would spray; in a household with ten cats, that likelihood jumped to 100 per-cent—there was at least one sprayer in every such bunch. It's some-thing to think about if you value the quality of life for all your pets—not to mention the smell and sanitation of your home.

Sometimes I see households with numerous cats having no problems whatsoever. If each has sufficient space to be separate from the others, enjoying his own special location where he feels com-fortable and can cope successfully with the stresses of daily living in his territory, there is less likely to be a marking or stress-induced urination problem.

But I've also seen big problems in households filled with cats. One of my first cases of too many cats took me to an old farmhouse on a dirt road in rural South Carolina. I found Louise Bell out back in the small barn, filling up bowls of food and water. A small army of dogs and cats milled around, waiting impatiently for her to finish.

"I sometimes feed a stray or two," Mrs. Bell explained, turning off a hose and wiping her hands on her apron.

"Looks like you've got quite a few takers," I remarked as the animals dived into the bowls of food.

"Oh," she laughed, "most of these are mine—my cats Lotus, Frenchie, and Rambo, and there's Rex—the cat I called you about—and over here are my dogs Daisy, Captain, and Brownie." She shrugged merrily. "The rest are some of their friends."

We turned toward the house. "Rex has been attacking Frenchie and also spraying all over," she began. She needn't have mentioned the spraying; a pungent odor wafted up from the damp earth. I took a deep breath before we entered the house and were enveloped in the all-too-familiar perfume of cat urine.

As I questioned Mrs. Bell, it soon became clear that Rex was doing his best to switch from being the "underdog" to head-of-household cat. The newest and youngest of the felines, Rex had spent his first two months hiding under the bed while the females stalked him.

"I felt sorry for him and started putting him outside a few months ago," Mrs. Bell recalled, "and as soon as I did, his personality completely changed. He started spraying and attacking the females, especially Frenchie.

"Frenchie used to be so affectionate," Mrs. Bell added sadly. "She would sit on my lap all the time. Now she just hides under the bed to avoid Rex. This has been going on since Thanksgiving." This was a good example of how a traumatic event can apparently alter a cat's personality.

It was now early spring, and Rex was becoming reluctant to go outside; when he did, there were frequent cat fights with the strays that hung around the backyard.

To top it all off, the cat had taken to grasping Mrs. Bell's arm, a sign to me that he was overly aroused and attempting to release his arousal by dominating or controlling anything he could get hold of.

So here was poor Rex, who had apparently taken it upon himself to make sure that he had control of everything outside and inside the house. If he couldn't control other cats' access to his resources and establish his space with spraying, he would do it with aggression and attacking, and—most unusually—"humping" on his owner's arm.

"Does Rex try to sit up high when he's in a room?" I asked Mrs. Bell.

"Why yes, he does," she replied, turning to the bookcases in the living room. "He sits up there on the top shelf whenever he can, or on the kitchen counter, or on top of the dresser in the bedroom. How did you know?"

"Well, one way cats have of indicating dominance is to pay a lot of attention to vertical spacing," I told her. "If you look at how cats arrange themselves in a room, the most assertive or the 'top cat' among them will usually be sitting up the highest."

I suggested she add a few more high shelf-type spaces for Rex around the house, as well as some intermediate and lower ones for the other cats, to facilitate a congenial social structure.

(Several years later I was intrigued to see a cat facility at a Florida animal-control shelter where, instead of living in individual cages, the cats had been allowed to arrange themselves on a series of planks up and down the walls of a room. I was struck by how calm and peaceful the animals appeared. The animal-control officer confirmed that before the shelves were constructed, there had been a great deal of crying and spitting and fighting among the cats. While large groups of cats may be more likely to spread disease if not individually housed, vertical spacing seems to be the cat's miaow.)

When Rex went outdoors, he marked his territory by spraying, I explained to his owner, and when he came back in, he and the other cats probably brought with them the scent of the strays. This led to more marking inside as well as attacks on his housemates.

"If Frenchie has the odor of the strays on her feet or hair, it may cause Rex to attack her as if she were the 'owner' of that urine," I said. "I wonder if you could keep Rex and the other cats inside more of the time, which should lessen the spraying." Mrs. Bell said she could try.

But I felt the situation would never significantly improve as long as the strays were allowed to remain in her yard. Although it was difficult for an animal lover such as Mrs. Bell to ignore a single homeless animal that wandered by, she reluctantly agreed to ask the local animal humane organization to take control of unowned visitors, so that the presence of the strays would cease to be a factor in her own household's problems.

The persistent urine odor in the house could perpetuate the problem, I told my client.

"I've been trying to keep it cleaned up," Mrs. Bell told me as she walked through the pantry and picked up a half-empty bottle of ammonia. "It's not easy; there's been spraying all over my bedroom and even in the kitchen sink." I recommended that she switch to a chlorine-based product or use rubbing alcohol or vinegar, explaining that uric acid breaks down into ammonia by-products, so it was possible that cleaning with ammonia was actually attracting the cats to that location for their urinating or marking.

"Oh, great!" Mrs. Bell replied with a wry smile. She tossed the ammonia into the kitchen wastebasket.

"You told me the cats don't play with one another," I said. "So now let's work on getting Rex and Frenchie to be friends. May I see the room where Frenchie hides under the bed?"

In her son's bedroom, I asked Mrs. Bell to try to tempt each of the two cats at different times to reach under the closed door with a paw to get at something on the other side—a playful behavior used by many felines. She could touch their paws herself or provide a favorite toy or a morsel of cat treat. Rex should reach in from the hall, Frenchie out under the same door from the safety of the bedroom.

Once that behavior was well established in both cats, I asked Mrs. Bell to see if the cats would begin reaching and touching each other's paws when they were both put in position on either side of the door. Before long each cat would see the other as something to play with, not something to fight with; the door would serve as a natural barrier until Frenchie's safety seemed certain.

I also suggested that Rex be fed and watered away from the other cats until he calmed down considerably. (He already had his own litter box, and now Mrs. Bell understood why four separate litter stations hadn't stopped the spraying problem, which wasn't an elimination issue.)

I recommended putting both Rex and Frenchie on the appropriate medication to lessen spraying and calm them down.

"Why Frenchie, too?" asked Mrs. Bell.

I explained that Frenchie's fearfulness may have been providing Rex with some sort of stimulus to attack her. "Applied animal behaviorists say that aggression is 'stimulus-dependent,'" I went on.

"If you are a cat, you don't attack a tree—you attack something that is moving or looking at you or coming toward you. There is a change in the cat's perception of that 'stimulus array' when drugs are administered.

"So while we'll be trying to calm Frenchie's fear, we'll also be trying to change whatever she's doing to elicit an attack from Rex."

"This is more complicated than I thought," Mrs. Bell said, shaking her head. But she agreed to follow my recommendations.

In our follow-up call a week later, Mrs. Bell reported that Rex's spraying and attacking had all but ceased. Unfortunately, in her enthusiasm to get that awful smell out of her house, Mrs. Bell went a step too far for her pets' sake. A month or so after what I thought was a final phone consultation, I got a frantic call on my answering machine saying that Rex had gone crazy, reverting to spraying all over the house. It seems that Mrs. Bell had called in workmen to take up the kitchen floor where the urine had soaked into the linoleum. This was too much for Rex; he couldn't control his space or the strangers in his house, and he suffered a major setback. But when the work was done, Mrs. Bell patiently started again, and gradually the problems lessened to a degree she could live with, along with her seven pets. She was quite a trouper.

One thing owners can do to help their cats get along is to rec-ognize that cats have different roles to play in relation to one another and to reinforce—not ignore or try to change—the roles the cats have chosen. That's not always easy.

Although words such as "jealousy" and "revenge" aren't gen-erally a part of my particular vocabulary—I try to work with observ-able patterns of relationships and other learned behaviors—I couldn't blame my client, Sally Butler, for describing her two-year-old male Binky as "the most jealous darned cat I've ever laid eyes on." After hearing about his exploits, I could see exactly what she meant.

Binky was making his "friend" Ruffian's life unbearable, Ms. Butler explained as we sat in the kitchen of her modest ranch house in an Atlanta suburb. She cradled Ruffian—a gray two-year-old male—in her arms as we talked. Over the next hour and a half, the cat stayed close to his owner's side or sat sheltered in her lap. He was obviously very dependent on his human friend.

"Binky won't even let Ruffian play with his own toys or use his own scratching post," Ms. Butler said indignantly. "When I pet Ruffian, Binky attacks him. He's even started attacking *me*." She had contacted me through her vet after Binky had bitten her twice in one day.

As if to confirm her tale of woe, the big white cat appeared and growled as Ms. Butler set the smaller gray one down. Binky's boldness in front of a stranger was impressive.

"No, no, you bad boy," she scolded, shaking a finger at Binky as Ruffian slinked off. "The last time I said 'no' to Binky, he bit my ankle," she confided unhappily.

"Well, let's see if I can give you some better tools to work with. Incidentally, do you try to pet Binky, as well as Ruffian?"

"Oh, yes," she replied, scooping up the bigger cat and setting him on her lap. "I'm very careful to give them both the same amount of attention. If I pet Ruffian, I pet Binky. If I give Binky a treat, I give Ruffian one, too."

She held the white cat in front of her face, nose-to-nose. "Mommy doesn't play favorites, does she, Binky?" she asked the cat proudly. He growled in reply, and she quickly set him down.

"I'm afraid your equal treatment may be part of the problem," I told her, then noted her obvious confusion.

After thirty minutes of diagnostic questioning, I had become fairly certain that Binky was the boss, and Ruffian the subordinate cat, and we were not going to try to change that.

"In fact," I said, "what you need to do over the next few weeks is to help reinforce each cat's role, with activities that define and separate the two of them. I suspect Ruffian's getting beat up because

Binky thinks he's too uppity, and giving Ruffian so much positive attention may be rubbing salt in the wound." Binky, I went on, had reacted with irritation to Ms. Butler's attempts to decrease the status difference between him and Ruffian, making the cat more likely to strike out at his owner. It was also not unusual for cats to punish the subordinate cat, thereby reestablishing the distance in status and stabilizing the role relationship.

"But it wouldn't be fair for me to ignore Ruffian!" Ms. Butler objected.

"Not ignore him," I rejoined. "Let's figure out some ways to make everybody happy."

Since Ruffian had reached the point of near paralysis with Binky monitoring his every move and threatening to attack, I told Ms. Butler to put Binky outside when she planned to have a play session with Ruffian. Then Ruffian could play without fear and she could pet him to her heart's delight.

"When both cats are in the house, you must feed and try to do everything with Binky first; let him play with Ruffian's toys without scolding him and so on. And you should let only Binky lie on the bed, not Ruffian. Binky needs to know that he is the 'preferred' cat."

"But—"

"Believe me," I assured her, "Ruffian is going to be much safer and more relaxed and Binky less 'jealous' if you stop trying to make them equal."

"All right," she reluctantly agreed. "Then will Binky stop biting me, too?"

"He should, but you can help with that, too."

I explained that aggression does not occur in a vacuum; before an attack takes place, there is always a stimulus that makes a cat respond. I'd noticed that Ms. Butler was frequently presenting Binky with her face, which made it very logical for him to lunge and bite her if he was already upset. A little distance during and just after play sessions would go a long way in solving that problem.

"He also bites my ankles when we're on our nightly walk," Ms. Butler offered. It seemed that the cat would walk with her down the sidewalk past ten houses or so, then turn around and walk home, just like a dog. No leash was needed; the couple just took a stroll in the neighborhood together, with the cat heeling at his human companion's side.

Unfortunately, when it was time to turn in at their own front walk, Binky always took the opportunity to lunge at his owner's ankle and sink his teeth in. It was obviously part of a play ritual the cat had built up.

So I asked Ms. Butler to invest in a number of cat toys and to carry one with her on the walk to throw as a distractor every night as she turned in at her own address. The rest she should keep in various locations in the house to head off imminent attacks on Ruffian or herself.

After my visit, Ms. Butler called me weekly with feedback, and in the latter half of the six weeks there were only two incidents—one in which Binky jumped Ruffian in the yard and another when Ruffian was inside (being petted in front of Binky, my client admitted guiltily). The white cat was even fetching the toy mouse its owner threw as they reached the driveway on their evening walk, rounding out its caninelike routine.

"It's been hard for me to treat the cats differently," Ms. Butler reported in her final phone call, "but it's worth it." She laughed. "I've finally got two cats that are separate but *un*equal, I guess, and now we're all getting along much better!"

Both cats and dogs show hierarchies of dominance and subordinance in relationships. However, cats are different from dogs in that the subordinate cat's behavior is motivated more by avoidance or defensiveness than it is by submissiveness. Complicating the situation in the multiple-cat household is the fact that as long as a cat's personality is not extremely fearful or extremely assertive, he can occupy a dominant role with one cat and a subordinate role with another cat at the same time. The rules that regulate their behavior are different with each role. The subordinate cat is likely to avoid confrontation over a desired resource rather than submit to the dominant cat. The behavior is more like timesharing of important resources than out-and-out competition for them.

Because of this behavior, it can be tricky to determine which is the dominant cat and which is the subordinate. One animal behaviorist, Penny Bernstein, used a cardboard box to test out the ranking she had come up with after observing a household of fourteen cats. She identified two as dominant. When she put the box down, both curious cats jumped right in, occupying it first—and more often—than did any of the other cats. (They did this at different times throughout the day, "timesharing"—not competing.) Owners can try this themselves if they have lots of cats and a good idea which one's on top.

Of course, cats' relationships to one another can change. Just when you think you've finally solved the mysteries of the feline hierarchy, those cats might throw you a curve.

One of my most puzzling cases involved an unexpected role reversal which took me a while to catch on to. It involved two female spayed domestic shorthairs: three-year-old Sleet and two-year-old Freckles. I met with their owners one autumn afternoon in an attempt to resolve a serious problem with inappropriate elimination and aggression.

Wherever Freckles slept, Ruth and Max Benjamin reported, Sleet would urinate. If the younger cat started to lie down on the owners' bed pillow, Sleet would rush over and urinate there. Freckles's favorite resting places—a chair, a hope chest in the bedroom, and the blankets at the foot of the bed—were also frequent targets. On top of this, Sleet was attacking Freckles three to four times a day. Not surprisingly, my initial notes show that Sleet was unquestionably the dominant one of the pair.

The family had moved within the past month—a red flag that often signals adjustment trouble for cats. The move had apparently disrupted the pattern of activity that stabilized the role relationship between Freckles and Sleet in the old house. Instead of reestablishing the old pattern in the new location, it looked as though the move had emboldened Freckles to try to take over the role of dominant cat; a formidable task best abandoned, in my experience.

Right after the move, Freckles had grabbed all the choicest places to lie on—including the comforter at the foot of the bed, a chair, and other locations where Sleet had formerly been able to rest. Now, whenever Sleet walked into the room, she first hid under the bed to see if Freckles was around. And even though she wouldn't challenge Freckles for these locations, when Freckles came away from them Sleet proceeded to urinate wherever the orange-and-white cat had been sitting.

This antisocial behavior didn't deter Freckles. She was persistent in choosing the best locations, and Sleet was beside herself with frustration. So, several times a day there was a chaotic chase, with Sleet pursuing Freckles around the house, Freckles running for her life, and the two of them wreaking havoc as they tore in and out of every room.

My first impression was one of the gray cat trying to maintain her dominance, and I thought the best thing to do was to encourage that. I didn't count on Freckles's persistence in trying to effect a role reversal.

So the treatment plan I formulated for the Benjamins was based on giving Sleet the best resources and attempting to reinforce her rapidly eroding dominant position. I told them to allow only Sleet access to the bedroom with all its preferred resting places, and have Freckles stay in the den at night, where the couple should play with her before bed.

I gave them some tips to minimize the urination problem and suggested they continue to greet Sleet first, feed her first, pet her first, and allow her to have higher resting locations during the day. Whenever possible, I told the Benjamins, they should encourage Freckles to lie in lower locations than where her companion was resting. With these not-so-subtle manipulations, each cat would remember her proper role and act accordingly. Or so I thought.

Wrong. After ten days Mrs. Benjamin called me to report that things weren't going quite according to plan. She had dutifully separated the cats the first three nights in a row, and this had seemingly had some effect on their behavior. The chasing had been drastically reduced.

"But Freckles doesn't like sleeping in the den," Mrs. Benjamin complained, "and we had to force Sleet to come in with us. We just couldn't coax her to come into our bedroom without Freckles! Isn't that odd?"

Very odd, for a so-called dominant cat.

So the soft-hearted owners had been letting both cats lie on the bed and encouraging them to get along—petting and talking to them. There had been no urination problem all week, and the cats were now "kissing" and licking one another on the mouth before going their separate ways. They were also lying together in the front window for the first time since the move. Mrs. Benjamin also reported that Freckles seemed determine to lie in Sleet's spots, no matter what.

And a new behavior had cropped up. The cats were getting into a ritual of jumping on the sink and drinking from the faucet before feeding. I explained that this is part of "chaining" behavior, which some cats start up to an hour before feeding. They may start the ritual by sitting on the bed while the owner dresses for work, then drink from the faucet, then sit in the window, then go to the refrigerator as the owner gets the food. Each ritual step in the chain leads to the next and the whole thing ends in breakfast. Mrs. Benjamin did not mind the drinking from the faucet, so I didn't offer any suggestions to interrupt this ritual. And I suggested very little intervention in the next week, as I wanted to see which way the cats were going to lean with respect to dominance.

A week later I got my answer. Things had gone pretty well in the previous seven days, with Freckles lying on the bed and Sleet under it most of the time. But then Mrs. Benjamin observed Sleet come into the bedroom and urinate on the cedar chest. She also told me that the cats had resumed chasing, but now the aggression was more "equal." Instead of Sleet terrorizing Freckles, Freckles was now chasing back.

At this point I had to admit that the roles were clearly reversing, and we'd better go with the flow. Explaining that I had underestimated the strength of Freckles's determination to come out on top, I instructed the Benjamins to now try to enable Freckles to become more dominant.

I suggested that Sleet now go into the den at night. Freckles should get everything first and should have unlimited access to the prized master bedroom.

"I'm afraid treating Sleet as number one might have put a strain on things," I explained to Mrs. Benjamin. "If we don't get further problems, we'll know we've made the right decision this time."

I was eager to hear what would happen; I had to wait only four days for the follow-up phone call this time. It was good news— sort of. The cats were doing better; there hadn't been a single incident. But Freckles was occasionally hissing at Sleet and leaping about.

"Have you made the bedroom off-limits to Sleet?" I asked Mrs. Benjamin.

"Not really," she confessed. "It doesn't seem fair. And then Sleet comes in in the morning and urinates on Freckles's sleeping spot, or jumps out at Freckles from under the bed." She sighed. "Maybe we should get a puppy to distract the cats."

I told Mrs. Benjamin that I believed this would only make the situation worse, adding to the cats' stress and possibly making drugs the only alternative. I preferred to keep encouraging one of the cats to be dominant. Although that seemed unfair to her, I explained again the concept that the cats start to accept their treatment as equitable in terms of their relationship. And that enables each one to know her role and relationship to the other, which reduces the stress of the unknown. She was really not being unfair when she acknowledged and facilitated the cats' role relationships.

I wished the Benjamins were able to make the bedroom off-limits to Sleet and let Freckles get some respite. The clients needed to be willing to make some life-style changes, not get a dog. Mrs. Benjamin said she understood and agreed to try to continue the program in a more determined way.

The following week was a good one, with only a couple of scraps. Freckles was rapidly assuming the role of top cat, occasionally jumping on Sleet from the windowsill, and Sleet was adjusting well to being deposed. Both cats were getting lots of attention and affection from their owners, but in a slightly more organized way than before we met.

I heard from Mrs. Benjamin on several more occasions, during which time the incidents became fewer and fewer. When we last talked, it was evident to everyone that there was a new head of the household—as far as the pets were concerned—and the felines and their owners were comfortable with it. Sleet and Freckles had indeed reversed roles, and difficult as the initial transition had been for all concerned, this mystery had a happy ending.

8

Inside Cats, Outside Agitators

One of the biggest decisions you'll make about your cat's life-style is whether to keep it indoors exclusively, outdoors exclusively, or a combination of the two. Any veterinarian will advise you to keep the cat indoors if you want him to live a long and healthy life. For outdoors lurk such deadly dangers as predatory animals, traffic, the various snares and traps of humankind and nature, mean kids, disease, other cats, and God knows what else.

On the other hand, barn cats have survived for generations and would never voluntarily give up their mouse-hunting life-styles. You have to decide what makes sense for you and your cat, with all the potential hazards in mind. (And even if he's always outdoors and not really a companion, remember that no one else will see to it that he is fed and inoculated on a yearly basis.)

Occasionally, I let my two college-town cats out briefly in the morning, before they're fed, for they love the sun and the fresh odors in the grass of my little backyard. But there have been times when one or the other has wandered next door, if only for a few minutes, and I wondered if I was doing the right thing. Creatures of habit, they have always returned safe and sound. Knock on wood.

When they are young, cats will adapt fairly easily to whichever mode you select. But as circumstances change, you may want to bring an outdoor cat in, or sometimes let an indoor cat out. As a cat ages and her sharp eyesight and hearing begin to go, she will no longer be able to ward off dangers as she did with all her faculties intact. Similarly, if the cat is declawed, her defense and escape potential might be compromised, although I've seen cats declawed on the front paws have no trouble climbing trees or bluffing successfully with a clawless swipe at a dog's nose. But whatever the reason, bringing the cat indoors after she has enjoyed her freedom for a while can be a challenge for both of you.

Hershey and Splash's days appeared to be numbered by the time I met their fed-up owners. These two old tomcats had had a lot of adventures to look back on, if cats do such a thing. They had spent their nights outdoors for years, prowling around the neighborhood and having a grand time.

But now, all that had changed. Some new cats and dogs had moved into the area, and there had been so many cat fights and narrow escapes that the moonlight marauders were finally persuaded by their owners (and the neighbors) to retire to the house at night.

"I thought they'd just curl up and sleep all night, like the dog," Paul Marino complained as I sat down with him one autumn evening. His canine companion, a mixed breed named Newton, snoozed at his feet. "But these cats are driving me crazy," he continued. "I haven't had a good night's sleep since I brought them in a month ago."

It seemed that Mr. Marino had been more than generous, even giving the cats their own room—one of the bedrooms in the eight-room colonial. His two kids had both gone off to college, so there was plenty of space for the felines.

"I tried separating them for a while, giving each one his own room, but they only got worse," the frustrated owner reported. He was getting increasingly upset as we talked. Probably the lack of sleep was making him cranky.

"We should have sent them off to the university with the girls," he grumbled. "I've got to get up and go to work in the morning." He glared at me with furrowed eyebrows, apparently waiting for an answer to his dilemma.

"So your cats are keeping you from a good night's sleep," I said, reiterating his complaint. "But you did the right thing to bring them in at night—it's a jungle out there."

My attempt at levity produced only a deeper scowl. I hastily dispensed with the small talk.

"I'd like to know exactly what the cats are doing that you want changed," I told him.

"Well, Doris and I would like to be able to sleep all night without having a couple of cats throwing themselves at the bedroom door at three A.M."

"OK." I made a note. "What else do they do?"

"Well, they race around the house crying and caterwauling— eeooow! eeooow!—and making such a racket I'd like to wring their necks!"

"Well, I'm glad you called me instead," I said weakly.

". . . always jumping up on the kitchen counters, knocking everything off, spilling the sugar—"

"Calm down, dear," interjected his wife, turning off the computer in their family room. "I'm sure Dr. Wright gets the picture." She peered at me over horn-rimmed glasses.

"Oh yes, I do, indeed," I replied hastily. "Now let's see what we can do about this situation."

"Good. I'll go call in the cats," Mr. Marino said, starting for the back door.

"That's OK," I stopped him. "I don't really need to see them at this point."

"Suit yourself," he replied stiffly, shooting his wife an incredulous look. I could see he was wondering how I was going to "cure" the cats without ever laying eyes on them.

I had spoken to his wife on the phone, and I now realized that she hadn't spent a lot of time sharing with her husband the information I had given her on how I went about my work.

"Short of camping out in your bedroom at night to see for myself, your description of your pets' problem behavior is the best tool I have to work with," I explained. "May I see the bedroom areas?"

"Sure, go on up," said Mrs. Marino, as she waited for her husband to join her. I went up to the second floor and waited on the landing as their voices floated up the staircase.

". . . high-priced quack—"

"Shhh!"

Finally we got started. The couple showed me the bedroom where the cats had been sleeping—they used the term loosely—and then led me to the master bedroom.

"I take it that letting them have access to your room at night would be unacceptable?" I asked. Sometimes the easiest solution completely escapes people.

Mr. Marino rolled his eyes. He clearly considered me a moron since I refused to examine the cats.

"You take it right," he said sarcastically.

"We did let them in here at first," his wife said pleasantly, throwing daggers at her spouse, "but they would get up in the night and race around the room and jump on the bed—"

"Damn near gave me a heart attack!"

"Don't excite yourself, dear. . . . Well, we just couldn't have them in here."

"So what did you do next?" I asked.

"Gave them their own room," Mr. Marino replied. "Pretty soft life for a couple of alley cats—you know, there's people sleeping on park benches downtown that would give their eyeteeth—"

"Hershey and Splash didn't take to being kept in a room at

night," Mrs. Marino cut in. "They threw themselves against the door, and they banged on the window screens and made quite a fuss until we let them out. We've tried putting them in various other rooms, but the same thing happens. Even if they're downstairs and we can't hear them, I'm afraid they're going to hurt themselves."

"So now they're banging on *your* door," I concluded.

"You got it, sonny," Mr. Marino replied. "Now, what are you going to do about it?"

"Well, I think there are some steps we can take to improve the situation," I replied, trying not to take his hostility personally. "But first, let me ask what you do when the cats bang on your door at night. Do you just ignore them, or do you do something else?"

"Ha! *You* try ignoring all that banging and crying and scratching—"

"We usually *try* to ignore it for fifteen minutes or so," said Mrs. Marino, "but by then Paul usually gets up and kicks on the door until they stop."

". . . give them a dose of their own medicine . . ." the man muttered.

"Or I go to the kitchen and get them something to eat," Mrs. Marino added. "That usually holds them until morning."

The disturbances these two cats created were magnified by their occurrence at night, which disrupted their owners' sleep. But I felt the behavior was basically a short-term, exaggerated response to the stresses of being kept indoors after years of roaming free at night. The best way to deal with the problem was to try to reduce the tension the cats were feeling and to somehow lessen the chances of their owners being wakened in the night as long as the problem behaviors continued.

I asked my clients to close the screened windows at night and to put up the storm windows as soon as the weather permitted. This would keep the sounds and smells of any outdoor animals outside and cut down on the cats' arousal in the middle of the night. I suspected that what they *really* wanted was to be back outdoors where the action was, and that some odor or sound was stimulating them to action.

Mrs. Marino was on the right track by placating Splash and Hershey with food in the night—but unfortunately the cats had trained her to get up and feed them at 3 A.M. Now their cries also meant "feed me!" in addition to "open this door!"

I asked Mrs. Marino never again to reward the cats' cries with food and suggested that she buy some cat treats or dry food to leave out at bedtime in a location away from their bedroom. This way the cats (whose nocturnal exercise over the years had kept them pretty lean) could nibble away without disturbing their owners or perpetuating bad habits.

It took some persuading, but I convinced Mr. Marino to attach a piece of acrylic plastic to the outside of his bedroom door, which was hollow and reverberated with each bump and scratch. This served to deaden any noise the cats continued to make at night. Because the Marinos could sleep through this muffled assault, the cats would begin to learn that those behaviors did not result in any kind of reward.

To deal with the jumping up problem, I taught the Marinos how to rig remote punishers—in this case upside-down set mousetraps—on any counters the cats frequented in the middle of the night. They decided to use kitchen towels hanging over the edge of the counters as signals that the mousetraps were there, instead of the waxed paper other clients have chosen. I leave it up to the client to select the tools for this exercise; usually something they already have in the house works just fine, as long as it is novel and distinctive.

I also asked the Marinos to praise the cats and play with them in the daytime, so that Splash and Hershey would begin to find the house a more pleasant place to reside.

The attic fan could be turned on to mask any crying that persisted (from inside or outside) as the cats adjusted to their indoor habitat. Thus, by removing the attractive cues from outside; eliminating the scratching and leaping behavior and noises with the Plexiglas, traps, and attic fan; providing the cats with food on demand; removing any attention they were getting for their troubles; and providing a more interesting home environment, the cats had a real chance.

The risk I took was that the cats might begin to reinforce their own misbehavior by eating immediately after performing each disturbing action. Fortunately, they did not.

I didn't hear from Mrs. Marino until a few weeks later. She reported that the cats seemed much more content, and the couple hadn't been awakened at night for a week. Then she told me to hold on. When she came back on, it was to deliver a message from her husband.

"He says to tell you that he knew the minute you stepped in the door that our troubles were over."

Right.

M r. Marino's cats were making him a bit cranky, but I can recall another client who was utterly exhausted from sleep deprivation, due to her cat's nocturnal activities. The young woman came to me with an indoor-outdoor cat problem that was robbing her of a good night's sleep every night. Carly Wilson's two-year-old female spayed cat, Simca, had made some friends and some enemies outside the apartment and wanted to join them at night for whatever cat activities were on the docket.

Then, having worked up an appetite, she wanted to come in and get a bite to eat. After a while she was ready to go back out again. And so on, and so forth, until dawn—when she contentedly went to sleep on the bed, just as her owner was dragging herself out of it.

Because of all the cats outside, the owner had no intention of installing a cat door. And she couldn't cope with electronic collars and wiring her apartment. So the way Simca got in and out was to jump up from the back patio onto the window screen and howl until Carly opened the window. If she was hungry, she would paw at Carly's mouth until the owner got out of bed and fed her. When Simca wanted out, she sat in the window well and cried until the screen was lifted by the bleary-eyed owner. These activities continued all night.

Worse yet, the cat, in a constant state of arousal because of the outside agitators, had begun urinating outside her box as well as spraying. Carly didn't want to keep the cat inside all the time, because Simca seemed to have at least one cat friend out there—a male named Captain. She saw them together quite a bit, usually lying side by side, and Carly didn't have the heart to break up this apparent friendship. But in the past two weeks, Simca had begun urinating on Carly's clothing and had just sprayed on a beautiful down comforter in the bedroom.

In desperation, Carly had left the cat out all night for several days before calling me, and she was afraid that Simca would end up getting hurt by other, not-so-friendly cats.

We tried all sorts of things to solve this problem, from putting up a sheet of plastic over the screen outside to minimize outside odors and sounds (the cat just banged on this barrier now) to setting a table under the window so Simca could look out all the time (but she still wanted to *go* out). I couldn't successfully treat the urination and spraying problem while the cat was still aroused much of the time. Under the circumstances, and with the owner's insistence on not using any medication to calm the cat, we weren't getting very far.

We decided to try letting Simca out only occasionally for short periods during the daylight hours before or after work when Carly was around to supervise and keeping the cat in throughout the night. I agreed that leaving Simca out at night could be dangerous, and the other felines didn't seem to be around as much during the day. The problem with this arrangement was that the cat continued her heart-rending wailing throughout the night.

My client finally produced the solution, imperfect though it was. Carly suggested that she could shut the cat in the large walk-in closet/bathroom area, where Simca's cries were so muffled by Carly's tons of clothing and the ceiling fan in the bedroom that the owner couldn't even hear them. The room was virtually soundproof. Considering our previous lack of success, this plan sounded like the most practical possibility left to try.

It didn't take Simca long to accept her new nighttime quarters. Carly put Simca's litter box, food, and water and a comfortable cat bed in the area, and just ignored the crying because she could barely hear it. It was a much less stimulating or arousing environment for the cat. Within a week, Simca stopped crying to go out because there were no longer any positive results from the noise, now that Carly didn't open the door or window for her.

Crying didn't get Simca out, crying didn't get her in. Crying had no consequence whatsoever in terms of Simca getting anything from Carly. If you don't respond to the crying by giving the cat something that she wants, then the crying will stop—or as behaviorists would say, the extinction of the response will take place. And with the in-and-out nighttime pattern broken, I was able to successfully address the spraying and urination problems.

What I didn't like very much about the path this case took was the period when the cat was forced to be unhappy, with the owner unable to do anything but totally ignore her to help the cat find a safer and saner solution to the nighttime problems. But Simca could have ended up being abandoned or euthanatized when the owner couldn't take it anymore. Sometimes you just have to do the best you can with what the cat or the owner can give you.

Simca adjusted pretty well, so I'll somewhat reluctantly call this case a success. The cats outside that were causing all the trouble went away; there was no longer any reason for them to come around. Simca was not all anxious and wrapped up in the other cats' goings-on anymore. She eventually stopped crying when her pleas were ignored, and the activity in the apartment was stabilized. Simca was able to go back and use the litter box regularly, and eventually come out of the closet.

If you decide to keep a former indoor-outdoor cat indoors exclusively, try to do it gradually if possible. Start by bringing the cat in an hour or two longer each day and increase the time by one more hour every day or so. Be prepared for your cat to sit at the door waiting endlessly for you to open it or wander aimlessly around the apartment crying piteously. You must be ready to distract the cat and try to make him feel better on these occasions. Play with him, give him a treat, or just set him on your lap for some extra petting. Make sure the cat has a number of vantage points to look out at his former stomping grounds. These types of activities and arrangements will help ease the transition. If you feel so sorry for him that you open the door and let him out, it will prolong the adjustment period. But it's up to you how long you want this to take.

If the cat genuinely seems to consider your home worse than Alcatraz, you might consider trying a harness and leash and taking him out a few times a day for a walk. Introduce the harness gradually, stroking the cat or offering treats as you place it around his torso. Not all cats will take to this. It may be an exercise in frustration if you don't desensitize him to the harness first—and he might spend the better part of the "walk" trying to slink or leap out of the contraption. But it's worth a try.

The harness can also be useful if you are trying to get an indoor cat acquainted with an outdoor patio. When you first take your cat outside, sit with her while she explores her new turf. Call frequently (if she responds to being called) to help her learn the appropriate boundaries of her territory. You can use the leash and harness as an extra precaution, especially for cats who don't come when called. (While a fenced-in yard serves to keep a dog confined, it is a useless barrier in the face of a feline determined to break out of confinement.)

Gender and sexuality can make a big difference to people with indoor-outdoor cats. If you intend to let cats outside, you might be interested in a recent study that gives us a good idea of typical feline home ranges, if they have the opportunity to travel as far as they want to each day. Tomcats—unneutered male cats usually in search of fertile females—ranged over an area equivalent to about two acres, whereas neutered cats of both sexes used about one-tenth the home range of intact males.

Another study tells us that feral (outdoor unowned) cats live an average of two to two-and-a-half years, while indoor cats now average seventeen years. That's something to think about before opening the back door.

Cats are generally more than willing to spend time outside. But a formerly sheltered indoor cat may not immediately have the street smarts to cope with all the dangers I outlined at the beginning of the chapter. So think twice about turning a cat loose outside without a pretty compelling reason. "Because she wants to" or "because we moved out of the city" don't necessarily fit the bill. If she is basically a homebody, she won't stray very far. Unaltered males and females in heat are another story. It's up to each one of us to assess the risks and make the best, most responsible decision for our cats.

9

Life-styles of Stressed-Out Felines

"You're right, guys, this was a bad one," I told my two cats as I set down my briefcase after calling on a stressed-out feline named Tipper. Instead of greeting me pleasantly as they usually did, my two pets had bristled when I walked into the house. Now Turk suspiciously sniffed at my briefcase, while Domino, hair standing on end, examined my shoes.

It was no wonder, my things were scented with urine. This poor cat I had just called upon had sprayed his owner's entire apartment. He had left his mark on the antique furniture, the living room couch and chair, a bookcase, the sound system, the bedrooms, and even the television. If there was any area he had missed, it also had escaped me.

The cat's owner, Stephanie Dunbar, had hoped that Tipper and his companion, a female named Foxy, would keep each other company and play together. And play they did. Unfortunately, Tipper was very high strung and romping with Foxy was so arousing that he had taken to relieving the stress by spraying.

In addition to spraying, Tipper had formed another habit to help himself cope. Not unlike an anxious toddler, Tipper had found some relief in sucking. Unfortunately, the only thing he was interested in sucking was his long-suffering owner's thumb.

Mrs. Dunbar had filled me in on her cat's unusual behavior an hour earlier. "It always starts with Tipper running to his water dish and taking several sips of water. When he does that, I can be pretty sure he's going to come looking for me."

"Is this usually after playing with Foxy?" I asked.

"Yes," she confirmed. "He's always pretty jumpy after they bat at each other for a while. That's when he either sprays or goes to his water dish."

"So he seeks you out . . ." I prompted.

"He seeks me out, gets my hand, and sort of grabs my thumb with his mouth—he never bites me—and just sucks on it like a baby!"

"And how long does this go on?"

"Lord! He'd do it for hours if I let him. I have to get my housework done. . . ."

I asked Mrs. Dunbar if the cat seemed very excited while he was sucking on her thumb.

"Oh, no," she replied. "It calms him down right away. He sort of goes into a trance and closes his eyes, his ears go back and forth as he sucks, that sort of thing."

"All right. Now what happens when you make him stop?"

"Well, that's another strange thing. He goes right back to the water dish, drinks some more water, then runs back to me and tries to suck my thumb again."

"That's very interesting," I said. "What Tipper's doing when he runs back to the water is called 'chaining' behavior. He's going through a ritual that he believes will lead to—and end in—his being gratified by sucking your thumb. If he drinks some water, what follows is your allowing him to suck."

Mrs. Dunbar looked at me curiously. "I'm not sure I get it," she said.

"Well, we would call this superstitious behavior," I told her. "It's pretty much the same thing as a major league ballplayer getting up the day of the game, always eating the same breakfast at the same restaurant, putting on the same dirty socks, and tapping home plate four times before his first turn at bat. He believes these things will lead to a winning game because the first time he did them, he hit a home run. Your cat is doing essentially the same thing."

Mrs. Dunbar shook her head as she rose from her odiferous sofa. "This surely is not what I had in mind when I took these two cats," she said dryly.

"I'll bet. Let's see what we can do to help Tipper cope with life in more acceptable ways."

I outlined a program for Mrs. Dunbar that included making the sprayed-on areas less attractive; holding, stroking, and gently playing with the cat each day, which she had not been doing; and temporarily dosing the cat with antianxiety medication.

"You might also try offering it a baby's pacifier," I added. "Then he could still comfort himself with sucking, and you'd give yourself a break."

On her follow-up call a week later, Mrs. Dunbar reported that Tipper had found a toy he liked—a Ping-Pong ball, which she'd begun throwing to him after he played with Foxy but before he sprayed, thereby breaking the established pattern. Now Tipper's energy was rechanneled into playing with the ball, followed by resting, instead of going on to spray.

With the medication, cleaning, and new play patterns in place, spraying incidents were reduced to less than one a week over the next several weeks and were continuing to abate at my client's last call. As for the pacifier, the cat had spit it out, just like a breast-fed infant accustomed to the real thing. So Mrs. Dunbar resigned herself to having Tipper suck her thumb until he gradually weaned himself from the need for this anxiety-reducing habit.

Stressed-out Tipper and his owner always come to mind when

I'm asked for an example of what cat people will put up with from the pets they love.

Mrs. Dunbar might have been even more resigned to her lot in life as human pacifier if she knew what some other cat owners had to endure with their needy pets. I remember one cat that went ballistic after being displaced by a baby, developing stress-induced seborrhea and vomiting, and starting to scratch all the furniture.

But his most desperate ploy came when he tried to solicit his owners' attention and they ignored him, preoccupied with warming the baby's bottle or changing diapers. When this happened, he bit their legs. Then they chased him, and he retaliated. This was his way of getting some much-needed attention, although it finally prompted his owners to call me in a panic. I had to help them understand their cat's need for companionship, and things rapidly improved.

Too much stress: It affects humans and other animals alike. With people, irritability and arguments might be the result; the victim might abuse drugs or develop ulcers or tension headaches, or blow a fuse and smash the contents of the china cabinet against the dining room wall to get some temporary relief. With cats, there's a good chance that too much stress will lead to urinating on the rug, crying, biting, chewing, or any number of other unwanted behaviors.

What exactly is stressful to a cat? It depends. As you know, cats have different personalities and preferences—and when the environment becomes chaotic or when strong preferences are thwarted, it can lead to trouble. It takes a special cat owner to see annoying misbehavior as a cry for help.

Carla and John Sinclair had just about had it by the time they called me. Their American shorthair, Cinder, was making their lives miserable, and they couldn't figure out why.

"She follows me around all day when I'm home, and goes crazy when I leave," Carla told me.

"She growls and hisses at all our friends," said John.

"She has accidents all over my side of the bed," Carla put in.

"She went number two in my suitcase," John added.

Cinder sat innocently by Carla's side, ignoring the testimony against her. True to form, she had growled at me as I entered the house.

"The vet tried giving her some medication, but she seemed to get even *more* aggressive," Carla informed me. "So that's when he recommended we call you. Why would a tranquilizer make her worse?"

"It's not unusual for that kind of drug to relax the cat's inhibitions against some kinds of aggression and allow it to act out even more than it did originally," I answered. "I can discuss medication with your vet if we decide to go that route. But first let's sit down and see if we can sort out the problems Cinder is dealing with."

The trouble started when the couple installed a sophisticated security system in the house, a modest ranch-type home in a neighborhood that had been increasingly plagued by burglaries. The system included a motion-detector that my clients wanted to activate in the main portion of the house when they left for work each morning.

After two false alarms set off by the cat, they realized that they would have to confine her to the bedroom while they were gone. Cinder had not adapted well to this predicament, and the rest was history.

"Before you installed the alarm, where did Cinder normally lie when you left and when you came home from work?" I asked the couple.

"She always sat at the living room window watching me drive away, and she'd meet me at the front door when I came home," Carla answered. "Sort of like a dog."

Based on the cat's distressed behavior, it was the answer I had expected—unfortunately. Unlike the many felines that couldn't care less what their owners were doing, this one cared a lot. It was the worst type of candidate for confinement.

"I'd like to take a look at the bedroom," I requested. We walked to the large master bedroom at the back of the house; Cinder trailed along, never leaving Carla's side. In the large, sunny bedroom, French doors led to a patio and the backyard. Neither the garage, the driveway, the street, nor the front yard was visible from there.

"Cinder is probably upset that she can't see you coming and going anymore," I told Carla. "She may perceive that it's part of her job to keep track of you. Does she object to staying in the bedroom when you leave in the morning?"

"Does she!" John burst out. "We have to chase her all over the house and slam the door before Carla sneaks out. Cinder cries and hisses at us, and she did this to me yesterday." He held out an arm covered with scratches for my inspection.

"And whenever anybody comes to the house, she stands in front of me and growls at them, the way she just did to you," Carla added.

Cinder was obviously very attached to Carla, I observed, and was exhibiting separation anxiety that had worsened since her owner had started shutting her in the bedroom. This problem was more common with dogs—the ones that tear up the place while their owners are away. I asked John and Carla what they found when they came home.

Plenty, it seemed. While the cat slept on John's side of the bed, she seemed to do nothing but go to the bathroom on Carla's side.

"And the litter box is right there, ten feet away!" Carla said with exasperation. "It's not as though she can't find it!"

"That's true," I replied, "but I see her food and water dish are right next to the litter box. Does she eat all right when she's shut in here?"

"No problem eating," John confirmed.

I explained that by placing the food and litter box close together they had probably set up some competing motivations for Cinder, as cats don't like to urinate and defecate in the same place as they eat.

It seemed she'd chosen to eat there and go to the bathroom somewhere else—like on the bed.

"Or in my suitcase," added John. "I travel a lot, and I'm sick and tired of finding these surprises in my bag before and after trips."

"Do you think you could keep the suitcase closed and put away until Cinder is retrained, or pack and unpack in another room?" I ventured. "The suitcase probably represents a litter box to her: It has four tall sides and she can get in and out of it; the box has great soiled clothing odors; she can move the material around with her paws, and so on. The cat may believe you're providing her with another toileting area, and she's just doing her thing."

After closing up John's bag and moving the food and water closer to the bedroom door, we went over to the bed. The couple told me that the cat's urine had sometimes soaked through the bedclothes.

Since they thought it was too risky to invest in a new mattress at this point, I advised them to clean everything thoroughly and to encase the old mattress in a zippered plastic cover to protect it and cut off any lingering odors that would attract the cat.

I then suggested that they make the preferred spot—the bedspread—inhospitable for elimination purposes.

"I'd like you to buy some plastic carpet runner and cover the bed with it when you leave for work," I told John. "You can just slide it under the bed when you get home. For the first week or so, spread a little creamy chlorine product on the runner."

"What for?" asked Carla, wrinkling her nose.

"Well, cats usually scratch around and sniff and so forth before they go, and if the preferred location all of a sudden doesn't feel right or smell right, they'll go somewhere else—presumably in a nice, clean, private, accessible litter box."

"We'll give it a try," said Carla. "But Dr. Wright, if Cinder likes me so much, why is she only going on my side of the bed? And why does she hiss and growl at me when I come home?"

"All of that probably has to do with arousal," I answered. "But tell me first how you react when you come home and find the bed soiled."

"We take her and rub her nose in it. We want to teach her not to do that."

"And if she hasn't made any messes?"

"Then I make a big fuss and tell her what a good girl she is."

"How does she respond to that?"

"Well, actually, she usually runs and hides under the bed no matter what I do."

"Carla," I told her, "your responses are perfectly logical, and I know you want to do the right thing. But in both cases, the result is that Cinder gets all riled up and even more stressed. Now, she simply can't connect your punishment with her crime, which occurs some time—maybe hours—before you get home; but when you greet her, she figures she's going to be in for it, based on past greetings. This is probably leading to her defensive growling and hissing and so forth."

"What should I do?"

"I know it'll be difficult, but just ignore the cat and clean up the mess. Don't punish her and don't be overly exuberant if she has been good. We've got to get her overall arousal level down, and this is the best way to start."

I also suggested that when the couple left in the morning, they take a cat treat and throw it into the bedroom, away from the bedroom door, and then sneak out. This would help draw the cat to the bedroom without chasing, and help establish leaving for work as a more passive, pleasant time for Cinder.

Every now and then, I told them, they should go in and out of the bedroom, closing the door and opening it a few minutes later, and also leave for work and come back in through the French doors in the bedroom.

"That way the cat won't be able to predict with absolute certainty that you're leaving every time you close the bedroom door," I explained. "It should help defuse the situation."

"OK, if I can figure out how to get the French doors open without setting off the alarm," John grumbled.

"The reason Cinder urinates on your side of the bed, Carla, is probably that she started going there for comfort, to sniff around, after you left for work. But her state of arousal was so high because of the problems we've discussed that she began to lose control and to urinate there."

I had one more suggestion for making the bed a less desirable place for Cinder to answer nature's call.

"Spend some time playing quietly with the cat on the bed after you've come home and removed the plastic," I instructed them. "Leave her cat toys on the bed while you're home. Let's try to get that area set up for playing, instead of for what she's doing now."

With all these changes in place, Cinder began to calm down, and the problem behaviors were drastically reduced. And while she continued to remain very attached to Carla, the black cat was at last able to relinquish her guard duties to the security system that had caused all the trouble in the first place.

What is the cat thinking or feeling inside to make us an important part of its life? We don't know to what extent cats have emotions like ours, but it is obvious that they become securely—or insecurely—attached to people, and that attachment can be one where they're a bit more dependent or less dependent. I don't think that means that they "love" you more or less, but it's about how they've come to understand their relationship to you.

Cinder wasn't alone in her attachment to her owner. Many cats are more devoted to their owners than to their turf. They can handle big changes like moving to a new house with their owner, but not being left behind. It's the absence of a particular person—not just any person in the household—that they have trouble coping with. Such was the case with a Siamese named Suki.

I received a long-distance phone call one day from a young woman who wondered if I could help her mother, here in Georgia, cope with the cat the caller had left behind several months before when she moved to Nebraska as a newlywed.

It seemed that she felt guilty that the cat was causing problems for her mother, who had enough problems as a widow living on her own.

"I don't know why Suki is having litter-box problems," Beth Hoffman said in a worried tone. "She was always a good cat. Now she's even bitten my mom a couple of times. The catsitter too."

Beth said she had planned to fly in for a month, and would I be able to meet with her and her mother and Suki?

When I sat down with them in the pleasant split-level home the following Sunday, I was surprised to find two cats present: Suki and a calico named Blossom—"mom's cat." Blossom was problem-free. Beth seemed as oblivious to Blossom as her mother was to Suki. Interestingly, the accidents had increased since Beth's arrival, the young woman reported, and the cat had initially greeted her old friend with dilated eyes and bristled fur. Beth's presence obviously had a great impact on the Siamese.

Since I always try to rule out obvious litter-box problems first, I wanted to make sure the box itself wasn't the culprit. Its location seemed to be OK.

"This box is never dirty," bragged the mother. "My Blossom sees to that."

"What do you mean?" I asked. I've met some pretty self-sufficient cats in my work but never one that could clean its own litter box.

"Why, she's a regular little gossip," beamed mom. "She tells me all about it if there's anything in the box each morning."

"That's interesting," I said. "By the way, how do you know that Blossom isn't having the accidents?"

The mother's eyes widened. "Oh no, Blossom isn't doing this," she asserted, gathering up the cat and stroking it vigorously. "She covers up her business when she's finished. That's what I'm cleaning out of the box."

"And Suki—?"

"Suki doesn't cover it up," she sniffed. "Anyway, she's been going on the floor every day after I come home from work and let the cats upstairs."

To solve this problem, I felt the woman needed to keep Suki in the downstairs area until the cat could refrain from having accidents or being aggressive. A smaller space would make it easier for her to organize her activities. We went over several steps designed to help Suki return to using the litter box full time and to attack acceptable items such as balls of string and cat toys, rather than her caregivers.

But during the follow-up phone calls (made by Beth) over the next couple of weeks it seemed little progress had been made.

"Suki is still having accidents upstairs," Beth reported in a discouraged tone. "I wish my mom didn't have to deal with this."

"Wait a minute," I asked. "Isn't she keeping Suki downstairs as we decided in the treatment plan?"

"Uh, well, she doesn't seem to be able to do that," Beth said, hesitantly. "You know, my mother grudgingly took Suki when I moved away. But Mom doesn't really seem to pay much attention to her. I've been in and out a lot visiting old friends, but I don't think she's following the treatment plan very carefully."

"All right, I appreciate your being frank with me. Let's talk again before you go home and see if things have improved any. If not, I'll review the program with your mother and see if there's any way to make it easier for her."

"At least Suki is sitting on my lap and purring again," Beth said before she hung up. "I've missed that."

I did speak to Beth one more time, just before she took off for the airport back to Nebraska.

"Dr. Wright, my mom gave me a great going-away present," she said in a rush. "Here's my ride—anyway, it's a great new cat carrier. Suki's going back with me. Thanks for everything!"

The phone clicked in my ear, and I had to smile. "*All right!* I exclaimed aloud. "Good for you."

For whatever reason, Beth had decided that she could now take her cat with her, and Mom could devote her full attention to Blossom: the only cat she could love. It was unfortunate that Beth's mother wasn't able to give Suki the few simple things she needed to be a well-behaved, contented cat, but placing the right cat with the right owner turned out to be the best "treatment plan" for these two women and their feline best friends.

·

10

"Don't Blame Me—I Have a Note from the Doctor"

My thoughts often turn to a client who not only put his cat's needs above his own but did so at the risk of his own survival. Gary was a new client who asked me to come to his urban townhouse, which was in the process of total renovation. Outside, a gleaming red Porsche stood in the driveway, the dealer's sticker still on the window.

Letting myseif in as Gary had instructed, I was struck by the elegance of his bleached-oak floors, hunter-green marble foyer and pristine new white kitchen. Construction debris still littered the hallways, but I could tell the place was going to be really something when it was done. I was envious.

"I'm in here," I heard my client call quietly. Yep, this guy must have hit the lottery, I thought as I made my way through the decorator living room toward the small den in the old, unrenovated part of the house. In this section, the musty smell of urine rose from the carpeting. A pale, tired-looking man in his mid-thirties sat at a cherry desk, his beautiful Persian cat on his lap. He rose slowly and shook my hand as the feline jumped down.

"Hi, Silver," I said, squatting to offer the cat my finger. Silver Star slunk under the desk.

"Great Porsche you've got out there," I said, to break the ice. Gary smiled. "Yeah, I love to take it out and hit 120 mph," he said. "I always wanted to do that, just—you know—for the thrill."

Risky business, I thought, but every man's dream.

But turning to the problem at hand, it seemed that Silver Star had been terribly stressed by all the construction work, would not set foot in the new parts of the house, and was urinating all over the den, where his owner was holed up while the work was being done.

"I have a friend who has agreed to adopt Silver, but only if he isn't having any more accidents," Gary explained as we walked down the hall. "The poor cat is so upset, I just have to get him back to normal."

So the man would rather give away his lovely cat than risk further accidents in his fancy new house. Well, that would seem reasonable to some people.

"Can you please help us?"

I looked up in surprise at the desperate note in his voice. We had paused in the doorway of his handsome new bedroom. Beside the bed hung an IV setup, and on the nightstand was some medical literature. The words HIV jumped out at me. Now it all fell into place—the man had AIDS and was trying to get the most out of life while he could still enjoy it. Suddenly I no longer thought of his Porsche and his awesome home in the same light.

But now I was alarmed. "Do you have someone coming in to clean, or are you mopping up after the cat yourself?" I asked, turning back to face him. He shrugged.

"There's no point in having the cleaning woman until the work crews are finished," he replied. "But I'm taking care of Silver Star's accidents and his litter box."

"I wonder if that's such a good idea," I countered, hoping he wouldn't consider my concern an invasion of privacy. "Is Silver Star declawed?"

"Oh, no, I wouldn't do that to him."

Great. I really had this one pegged wrong. The man was actually jeopardizing his life for the comfort of his cat. With his severely weakened immune system, Gary was at risk of infection from cat scratches or bacteria in the urine or feces. Every prenatal class teaches this to pregnant women, but many other people—even some of those with AIDS like Gary—apparently do not know the dangers of toxoplasmosis. Or did he? Maybe holding and playing with Silver—his closest companion—gave him the same pleasure as racing his new Porsche at 120 mph down the highway. Dangerous, but that was what life was all about, when the time left for such experiences is short.

I told Gary about an organization based in San Francisco called PAWS that deals with dogs and cats and educates their owners concerning all the zoonotic diseases that can be transmitted to persons with weakened immune systems. Volunteers go to the homes of AIDS patients and take care of their pets, cleaning litter boxes and so forth, so that the risk to the owners is minimized, and they can continue to keep the companion animals they love. I suspected there might be a similar group in Atlanta, and Gary agreed to turn cleanup duties over to a friend or a volunteer, if he could find one.

"I guess I shouldn't be doing it myself," he confessed. One more thing to worry about, his shrug seemed to say. I didn't have the heart to suggest declawing now.

"Please watch out for Silver's claws," I said instead.

I set out a behavior-treatment program for him to follow in order to get Silver Star back to using his litter box. Things seemed to be going pretty well when I received Gary's follow-up call a week later, and he was getting some help with cleanups. I gave him a few additional suggestions, and he said he'd keep in touch with progress reports over the next couple of weeks. I was really glad he was able to keep Silver Star at his side. I never heard from him again.

Gary's situation was an unusual one to encounter in the course of my practice, in that a physically healthy cat was actually a threat to his owner's health. The overwhelming majority of cases in which my work involves a medical connection have to do with a physical condition or medical problem that has affected the cat's behavior.

For example, some cats become hyperactive or groom themselves excessively. These conditions may be managed by topical treatments, drugs, or hormones, and cats who are forever licking themselves can be given toys to chew or bite on. That's also the case with irritated skin causing tail biting. The cat behavior therapist will have to help the cat substitute desirable behavior ("chew on this toy") for the inappropriate pattern ("chew on my tail") the feline has built up during the course of the problem. Then the problem behavior may lessen gradually—or in some instances, stop overnight. Or, it may continue in spite of the resolution of the physical problem.

Unfortunately, owners don't always realize that the cat's misbehavior can be aggravated or even caused by a problem with its physical health. If the cat is jumping up in the air, scratching excessively, or running around wildly in the house, the first thing you want to do is schedule an examination with your veterinarian so that if there is some physical cause it can be identified and treated. Often, the vet's treatment of the problem will take care of the behavior problem.

Cats who avoid walking on the carpeting or who hide in the fireplace eighteen hours a day because it's the only place that's hard and cool should be examined for fleas—they are trying to tell you something. I knew a cat who would jump from table to desk to TV to chair to piano, all around the house, to avoid stepping on flea-infested carpeting. Although the owner knew there were fleas and the leaping behavior was weird, he hadn't put two and two together. Fleas can lead to lesions, scratching, and infection. And they cause irritability, discomfort, and stress, which may lead to aggressive behavior.

Suppose your cat is in your lap and she's in pain because something hurts—inflammation of a joint or sensitive skin from an allergy or fleas. When you start to pet her, she may associate you with the source of the pain and lash out to scratch or bite you. When the problem is cleared up, you may be left with a cat that continues to bite you—not because you still cause her to hurt, but because you did in the past and the cat isn't willing to try you again. (Would you place your hand on the stove again to see if it still sears your skin?) Or, if you're lucky, the cat will go back to being its formerly sweet-natured self. That's what happened with a kitten I met some years ago.

Bingo was only three months old, and she was a mess. I was called by her owner, a middle-aged man, because the kitten had attacked the vet during a recent examination and had also stopped using her litter box in favor of the carpeting. As James Reingold told me about the kitten's background, it seemed curious he hadn't figured out the cause of the kitten's misbehavior.

Adopted from a shelter at seven weeks of age, Bingo had been cleaned up, declawed, and box trained. But some problems followed her home. She had a digestive disorder for which the vet had prescribed daily medication. She had ear mites, requiring ear drops twice a day. Then she got ringworm and, to add insult to injury, had to have a bath once a week—sheer horror for this water-hating, self-cleaning feline (although I've known some cats to enjoy swimming).

After several trips to the vet, Bingo understandably had come to the end of her rope and let the DVM have it, scratching and biting her. Now, this was not a vicious animal nor one that was play-fighting; she simply was hurting and uncomfortable and wanted to be left alone. I devised a corrective program for the litter-box problem, but as for treating the aggression directly, we left it alone.

As I expected, this kitten's "anger" dissolved as her health improved and all the vexing procedures wound down. I had to hand it to James for sticking with his little shelter cat through thick and thin. After a great deal of time, effort, and money spent on her behalf, he was rewarded with a healthy, good-natured companion animal—which is really the outcome he was looking for in asking for help from the veterinarian and cat behaviorist.

Bingo's was a pretty straightforward and obvious case of a medical condition creating a behavior problem. The diagnosis for Scooter, a five-year-old Himalayan, was a lot trickier. To all appearances, Scooter's problems stemmed from a rather unpleasant temperament. She was high strung and destructive at night—she would race around the house banging into doors, tearing up the carpeting, dumping over the flowerpots, and scratching the sofas. And when her owner went on vacation, the cat attacked and bit the cat-sitter, who promptly turned around and charged my client, Vera Wallace, twice the agreed-upon fee for her trouble.

Now, certain temperaments, if they're extreme, can limit the effectiveness of treatment programs. For example, if a cat has an aggressive temperament, it probably bites and attacks people and cats even during play, making it difficult to treat any specific biting problem.

"The only thing that calms Scooter down," Mrs. Wallace told me wearily as I took down some background information at her home, "is petting her for half an hour or so, at four o'clock in the morning. It's only a week since I got back from my vacation in Mexico, and I'm exhausted already." She smiled feebly. "Scooter wasn't always like this, Dr. Wright. What did I do?"

"Well, you may have done nothing that contributed to the problem," I said, "but let me have a little bit more information about her first." I asked Mrs. Wallace what steps she had taken—besides the petting, which taught the cat how to get attention in the middle of the night—to try to get Scooter to behave herself.

I generally ask this kind of question to discover what kind of inadvertent damage the well-meaning owner will have to undo before we can treat the problem. Unfortunately, it turned out that Mrs. Wallace did have a couple of tricks up her sleeve. She had tried squirting her pet with a water bottle, which only made Scooter madder. She then tried tapping the cat on the nose, whereupon Scooter bit her. Now into the second week of the "night terrors," Mrs. Wallace had run out of ideas. I was thankful, on Scooter's behalf.

"If I shut her out of a room," my client told me, "Scooter gets absolutely frantic. She goes around slamming into the doors, and she's already ripped up the carpeting and destroyed her favorite toy. I'm really puzzled," Mrs. Wallace concluded. "She never had these spells before." At the same time, the vet was treating the cat for vomiting.

I asked Mrs. Wallace to first stop all physical punishment, then I told her about some things she could do to help defuse the cat's apparent anger. Among these were establishing a quiet zone, tossing toys when the cat became overly aroused in order to redirect her energy toward play, and as frequent holding and petting as the cat would permit, reinforced with quiet praise. These steps were helpful, and the aggressive behavior stopped for several days but then started up again for no apparent reason.

Finally, because Scooter was still vomiting, the veterinarian did a complete workup on the cat and found a fairly rare condition that we both agreed was actually responsible for her bizarre behavior.

In this case, a physical cause had never been completely ruled out—a procedure most of us recommend—before turning to a certified behaviorist. But because the vet and I had kept in close contact, we were able to piece the puzzle together. The condition proved to be treatable, Scooter was soon back to normal, and her owner got her first good night's sleep since she checked out of the Acapulco Princess.

A typical example I run across in which a behavior problem stems from a physical one is that of the cat who eliminates all over the house. The first thing I recommend to the owner of a cat with this problem is to have the vet check the animal. The vet may discover a viral or fungal infection, a bacterial infection, renal failure, or a disease such as diabetes. Or the cat might have a neuromuscular disorder that renders him unable to maintain control over his sphincter muscles during elimination.

Urinary-tract infections and other maladies including constipation and diarrhea can—understandably—cause the cat pain or discomfort. Colleagues Borchelt and Victoria Voith theorize that the onset of the behavioral problem may occur when the cat is using the litter box, experiences pain because of the medical or physical problem, and tries to run away from the pain. Often there will be an unpleasant trail of "evidence" to clean up following such an episode. When I encounter this, I suspect that the litter box has begun to represent pain and is becoming the place for the cat to avoid in the future.

Sometimes a one-time association of pain with the litter box is enough to turn the cat away indefinitely. We call this one-trial learning, which is common in learning to avoid a noxious or aversive stimulus. Even though the cat had been using the litter box for three years and the painful episode occurred only once, the cat is not willing to go back and try the box again, even after the infection has been cleared up.

Another possibility is that the cat will establish a preference for going in a different location that becomes difficult to change when the physical problem no longer exists. Unfortunately, the new preferred location tends to be on the bed or the floor somewhere rather than in a litter box. So people try to cope with this by moving the litter box to the new location, figuring that will take care of it.

The cat then usually goes on the floor next to the litter box or picks yet another new location in a different part of the house, because it is still the litter box the cat is avoiding.

The way we deal with this situation is to change as many aspects of the litter box itself as possible. (This will probably not be effective as long as the medical problem or pain persists.) Change the smell of the litter from chlorophyll to baby powder or make it odorless. Change the texture from coarse (clay) to fine (sandlike). Change the shape, size, and color of the litter box. Change it from lidded to no lid. The more changes, the less likely the cat will associate the new box with the pain-related urination or defecation.

If the cat also has learned to avoid the new box's location, another should be found. As you can imagine, all this change might take a long time for certain neophobic cats to adjust to, but it's your best bet, so some patience will be necessary.

Although he had no problems with urinating in the litter box, Biscuit, a male three-year-old domestic shorthair, was no longer defecating in his pan after a bout of colitis. Biscuit was a shy, jittery cat who overreacted to outside stimuli. Any crash or backfire or thing that went bump in the night would cause him to jump and run around and hide. In fact, I just caught sight of the tip of his tail disappearing down the basement stairs as I came in to see his owner, Charlotte Curtis.

At the same time, Biscuit was surprisingly playful. The large, tawny cat was reputed to enjoy chasing crickets and reappeared from the basement to bat around a number of cat toys during my visit. Usually, skittish cats like Biscuit don't play much because they are too nervous to concentrate on having fun. Or they are so busy looking around for hazards that they don't have the time to play. But this cat was pretty enthusiastic about his toys.

Biscuit was making his mistakes on the living room rug, in the study, and in the bedroom. His owner described how he would defecate, run, stop, then run across the hardwood floor, defecating as he went. It was pretty obvious to me that Biscuit was trying different spots and always ended up running from the pain. But of course no location felt good to him, least of all his box—where he would have first noticed the pain. When he did try the box, Ms. Curtis told me, he howled, jumped out, and ran to another location to continue.

Colitis can be associated with chronic diarrhea. You can imagine how upset the owner was to come home each day and see these streaks across her floors and rugs.

What we did was very straightforward. The litter box was in the bathroom and the feeding station was in the bedroom, so there was no problem with the setup. Ms. Curtis got two new litter boxes—different sizes, shapes, colors. We tried litter with chlorophyll. Biscuit seemed to like the smell; many cats hate it. We placed one box in the living room (likely to make Biscuit happy) and one in the bathroom (more to Ms. Curtis's liking).

I then asked Ms. Curtis to place a baby gate at the entrance of the bedroom (which adjoined the bathroom where his box was located) and another in the living room. Because the cat wasn't much of a jumper, we knew these would work to restrict his access to the rest of the house. At the same time, the feline would still be able to observe what was going on beyond the restricted area and could hear and see his owner. This setup encouraged Biscuit to pay attention to the location of his boxes and increased the probability that he would use them. It proved to be a successful strategy.

We put his food and water and toys in with him, so he had plenty to do, and he started going in the boxes again. So with the cessation of the colitis problem, combined with a low-ash diet to prevent it from coming back and a training program aimed at using the new boxes, Biscuit was back in business. Eventually he was again able to have access to the entire house.

Sparky, a calico, was another cat with a medical problem who had abandoned her box. Sparky had been a so-called model pet before she came down with Feline Urinary Syndrome, a medical diagnosis that can encompass a number of different maladies. The cat had been treated with a special diet and medication, but unfortunately there was no pill that would send her back to her box to urinate once her medical problems set in.

The owner called me after two months of living with the problem. She had had to replace a sofa because the old one was covered with urine. What happened next was the last straw. The cat had actually urinated on her owner's lap while they were sitting on the new sofa. This proved to be only a one-time event, but it was enough to prompt Elizabeth Hunter to call for help. In the meantime, Sparky was urinating on the linoleum, on countertops, on the dining room table, on windowsills, and on the cedar chest in the bedroom.

The owner's frequent out-of-town business trips and constantly changing work schedule seemed to also be working against us. Cats like things to be predictable, familiar, and controllable; a combination of the medical problem and Elizabeth's career moves had upset Sparky.

Like Biscuit, Sparky acted as if she were trying to outrun her pain. As we experimented with various different litter materials, the cat spent her time going to the box, urinating very quickly, and then zooming away, as if to say, "If I get out of here fast, maybe it won't hurt!" The fact that she continued to use the box to defecate, with absolutely no change in that routine, seemed to confirm that the urinary problem was the culprit in the daily house-soiling episodes.

After her most recent business trip, Elizabeth found that Sparky had gone in no fewer than fifteen different locations in the tidy little ranch house. Elizabeth was a very neat, almost compulsively well-organized attorney, and it was tearing her up to come home to the messes Sparky left behind. But her compulsiveness worked in our favor, as she followed my recommendations perfectly and kept very good records of what was going on with the cat. This was quite helpful in tracking Sparky's progress from week to week and figuring out what worked for this cat and what didn't.

We used the plastic carpet runners to cover the most frequently soiled areas after their daily cleaning, because the different feel of the surface was likely to make them less attractive for elimination purposes. Sparky was very playful, so we put her toys all over the areas that had previously been used for urination and I instructed Elizabeth to play with her four or five times a day in those locations for a few minutes each.

This proved to be an extremely successful tactic, and with her toys on the new sofa, she quickly changed it to a play area and never urinated there again. I had Elizabeth try to involve the cat in play activities that were the same every day, in the same locations, followed by petting and feeding at the same time as well. I wanted the cat to be placed on a very stable routine and to rediscover the litter box, which we changed in as many ways as possible.

Breaking the carpet-soiling habit took a little longer. After a rough first week and a half, the cat started to get the hang of using the box again for urination as well as for defecation. Within a month, things were going well enough with Sparky that Elizabeth was able to invite business associates over for dinner, something that had been out of the question for some time. And finally, fun-loving Sparky was out of the doghouse and back in her owner's good graces.

Sushi, an eight-year-old spayed mixed breed, had an unfortunate experience with declawing, which is normally a pretty routine procedure. Because the surgery caused hemorrhaging, she was left with very tender paws afterward—too tender to withstand scratching in her clay litter in the litter box. After the operation, she chose to take her business elsewhere—to the soft carpeting she could scratch on without experiencing pain.

First, I put a piece of carpet under Sushi's litter box and surrounded it with plastic carpet runners so that the only soft place to go was near the box. Then, since the cat had been trained on newspapers and other smooth surfaces, we emptied the box of clay and put in a few newspaper strips and some soft, sandy litter. Every other day we added some more litter, and gradually she was able to return to the box full time as her pain subsided.

Incidentally, there is no scientific evidence at this point that implicates declawing in long-term cat behavior problems such as urination or aggression. Declawing has received a lot of bad press in recent years because it's a painful procedure, and I can't say I favor it, especially if the sole reason is to prevent the cat from scratching household possessions. The suggestions in Chapter Five on teaching a cat good behavior should be tried instead.

On the other hand, I don't agree with the people who label all owners who declaw cats sadistic, selfish beasts. Given the high rate of infection from cat scratches, there is a case to be made for declawing if toddlers are likely to be scratched in defense by a kitten or cat with claws. Supervising young children and teaching them to be as gentle as possible with their pets can help bring both cat and child through this difficult toddler period unscathed and, in the cat's case, with claws intact.

11

When Bad Things Happen to Good Cats: Getting Past Traumatic Events

Suppose your cat is calm, friendly, and well behaved. She takes everything in stride, from two cross-country moves in two years to the arrival of a squalling baby to the barking dog that has taken up residence in the backyard. Then one day the kids rush out of the house, and the door slams on the cat's tail. She lets out a terrified yowl, runs upstairs, and hides under the bed for the next couple of hours. When you can finally coax her out, you see that the tail is uninjured, but your pet is not yet acting herself. So you leave her alone. Eventually she ventures out of the bedroom, goes back (cautiously) to the kitchen for a snack, and thenceforth stays away from the back door when the kids are around.

No harm done. Looks as if she even learned something to keep her from getting in trouble in the future. And that's what happens day in and day out to cats all over the world. Little accidents happen, the cat is OK, and life goes on.

But in some cases, a cat doesn't come out from under the bed. She refuses to go near the kitchen anymore, which is a problem if she eats and has her litter box there. Or she turns on her littermate or owner, if they happened to be trapped in the same upsetting situation. Or she starts having accidents. The cat has been traumatized, and it's time to intervene.

One of my most interesting cases involved a freak accident that had two Siamese cats and their owner in a tizzy. I received a call one night from a woman who was nearly hysterical.

"Oh dear! *No*, Mr. Wong! Stop that! Oh, my goodness. Now, Dr. Wright, I really need your help—Good heavens, Ling-Ling, that's enough! The poor dears! What will I do? I can't seem to—Oh, goodness gracious! No, no, no!"

"Try to stay calm, Miss O'Toole," I broke in, as the gasping from my phone grew louder. The elderly caller, Amy O'Toole, wanted to engage my services at the earliest possible opportunity. It seems that her two formerly docile Siamese cats had been "dreadfully upset"—and from the sound of things, their distraught owner wasn't doing so well, either.

"Don't try to separate them yet," I told her, "or you're going to get bitten or scratched. After they stop fighting, see if you can put them in different rooms."

"I'll try. Lord-a-mercy!"

"Be careful handling them because they're likely to be unpredictable for a couple of hours," I added, glancing at my appointment calendar. "I'll meet with you first thing in the morning."

The next day my distraught client opened the door of her tidy bungalow and out poured her tale of woe. Everything had been wonderful until The Incident. The cats had been best friends all their lives. They played happily and slept in the same bed. Cats and owner had all lived together in perfect harmony. Now their lives were a total nightmare.

The poor woman was beside herself. It was obvious that Miss O'Toole was a devoted cat lover whose life revolved around her beloved feline companions. The three of them had enjoyed a stable, pleasant, and predictable routine in their neat little house. Unfortunately, a freak accident a few weeks before had turned their world upside down in less than a minute. Now it was up to me to fix it.

Mr. Wong had been snoozing on top of a curio cabinet one afternoon, when his pal Ling-Ling had jumped up beside him. Startled out of his sleep, the male leaped up and landed on an adjoining staircase, where he managed to become entangled in a plastic grocery bag. Then, screeching and wailing and trapped in the bag, he proceeded to wedge himself in between the bars of the staircase.

Miss O'Toole had quickly rescued her terrified pet and no physical harm had been done, but the animal had obviously been traumatized. From then on it associated ("blamed" pronounced the owner) the unfortunate Ling-Ling with the frightening chain of events. Mr. Wong's playing gave way to attacking his former friend every chance he got (for "revenge," Miss O'Toole reasoned).

The cats' behavior now followed a predictable pattern. The male would begin stalking the female, and when he got within ten feet of her, his eyes dilated, he bristled, and he began to growl. Naturally, Ling-Ling would growl back and run, which in turn prompted Mr. Wong to attack her. This scene was repeated many times a day.

During the night, Miss O'Toole was awakened to the sounds of her "poor dears" growling and hissing as they mixed it up. The female cat was so upset that, while being comforted, she had even nipped Miss O'Toole, an unthinkable act in this former peaceable kingdom.

By the time Miss O'Toole called me, the once assertive and dominant female cat was spending her days crouched beneath a chair in the sunroom, hiding from her stalker—who prowled nervously around the house. As for Miss O'Toole, I had rarely encountered a more distressed client. But I could see she was very wary of making *any* kind of change that might affect her cats, so I had to take her own anxiety into account when proposing a course of treatment.

"I feel the cats could both benefit from some short-term drug intervention while we work on their problem," I ventured. Her eyes widened.

"Drugs? What kind of drugs?"

"Well, both tranquilizers and female hormones have a calming effect on cats while they are learning new behaviors," I explained. "If you'll consider doing this for your cats, I'll discuss it with your veterinarian, and he can answer any other questions you might have about the particular medications before he prescribes them."

I usually like to wait a week or two to see if the drug route is absolutely necessary. But clearly there was no time to lose in Ling-Ling's and Mr. Wong's case. The longer we waited, the better chance the cats had of establishing a habit of reacting to one another negatively and the less we could expect to return them to normal.

"You're not going to turn the poor dears into a couple of limp dishrags, are you?" she asked anxiously.

"Not at all," I assured her, "and I understand your concern. This part of the treatment program will just make it a bit easier for your cats to get back to normal." A tranquilizer wouldn't have done *her* any harm either, it seemed to me, but that wasn't my department. Miss O'Toole reluctantly agreed to discuss medication with the vet, and we turned to the behavioral portion of the treatment program.

First, we went over a couple of tools she already had at her disposal to distract Mr. Wong when he began stalking. He sometimes responded to a sharp "*no*," and he would often jump up on her lap when she called "Where's my boy?" I urged her to continue using these distractors before the cat got to the growling stage. If he did stop stalking, she was to praise him for his good behavior. The idea is to train the cat to omit the attack and do something else; any other behavior will do. So at the first sign of stalking, Miss O'Toole learned to give him a signal that would elicit a change in behavior from aggression to anything else—coming toward her, playing with a ball, going to the kitchen to eat, or whatever. This procedure is a form of "omission training."

Ling-Ling needed to have freedom of movement again instead of staying frozen in terror beneath the chair much of the time. We arranged for Miss O'Toole to get a small harness and leash for Mr. Wong and to put him out on the back deck a few times a day, as she didn't feel comfortable letting her indoor cat run around loose outdoors. He (and his owner) made a good adjustment to this new temporary routine. And while Mr. Wong was on the deck, Ling-Ling began to learn to move around the house without fear and start playing again with her toys and with Miss O'Toole.

At night, the cats had taken to roaming around the house, suddenly coming face to face and scaring the hell out of one another—followed by nighttime hostilities. It was going to be necessary to keep the cats apart in separate rooms, a tactic I encouraged as much for their owner's sake as for their own. The cessation of predawn brawls couldn't help but improve the situation, and to ignore them would have defeated the efforts and gains realized during the day.

But the core of the program, and its success, depended on the cats' ability to become less fearful in one another's presence. With their daily (and nightly) contact more carefully controlled, we could now concentrate on exposing the cats to each other in a structured, non-threatening way.

I asked Miss O'Toole to buy or borrow two kitty carriers, and we set them at either end of her living room, about fifteen feet apart. The cats were facing one another but were separated widely enough that they didn't actually notice each other.

These exposures took place three or four times a day, for about thirty minutes each—or as long as Miss O'Toole was able to devote to the trial. Then the cats were separated in one of the ways we had worked out.

Each day, the cages were moved a little closer together. I cautioned Miss O'Toole not to bring the cats too close too quickly, as this could worsen the problem. If either cat reacted fearfully to the other—with growling, dilated eyes, fur standing on end, and the

like (an indication we were moving too fast), they were moved far-
ther apart. I also had Miss O'Toole place occasional food treats in the
cages to enable the cats to feel good in one another's company. Gradually each cat came to see that there was nothing to fear from the
other's mere presence and that being close to one another didn't
cause them any pain.

With the medication helping to relax them, the cats were able
to make rapid progress. Incidents became less frequent, and Miss
O'Toole's follow-up telephone calls changed dramatically from that
desperate first one.

"Dr. Wright, nothing's happening."

"Good."

"Is that all right?"

"It's perfect. You're doing a terrific job."

"Oh, OK!"

After about a month, the cats were off the medication and
starting to play together; Ling-Ling had reassumed her dominant role,
and the follow-up phone calls ceased. But I heard from Miss O'Toole
one more time, about a year later. She just wanted me to know that
life was worth living again, everything was better than ever, and
that she and her pets had never been happier.

Fear also played a major role in creating problems for Ginger, a cute shorthaired domestic female just over a year
old. The cat was the victim of a traumatic accident that constantly
came back to haunt her. The client called on me for help after several episodes of being awakened at three o'clock in the morning by
his cat, who he described as aggressive to himself and others.

When I took down the cat's history, it was easy to trace the
problem behavior to its root: a long car trip several months earlier
when Ralph Lewis and his cat were moving from the Southwest to
the East Coast. Ginger had been given the free run of the car. (Error

number one.) Although Mr. Lewis had given her a commonly prescribed sedative for the trip (error number two, wrong drug for a free-
ranging cat), the cat was exploring rather than sleeping. Nevertheless,
everything was fine until she crept under the car's pedals and Mr.
Lewis started to apply the brakes.

A terrible melee ensued, in which Ginger, hurt and scared,
latched onto her owner with teeth and claws, scratching and biting
him severely. The best that can be said for the whole episode is that
the owner was able to pull the car over to the shoulder of the highway without killing himself and his cat. Beyond that, his days of traveling with a carefree cat were over, at least for the time being.

Thereafter, whenever Mr. Lewis had to put Ginger in the car
(he took frequent long trips), the cat became aggressive and upset,
and my client learned to his dismay that he couldn't even touch
the cat after he closed the car door. Now, three months later, Ginger was attacking Mr. Lewis and others in the house as well as in the
car.

When I explained the course of fear-related aggression to him,
Mr. Lewis told me he usually could tell when his cat was getting upset.
Ginger was a cat whose whiskers, eyes, ears, and fur could be used
to predict when an attack was about to take place. Some cats show
one or two of these physical indicators, but Ginger showed them all.
Her ears would flatten against her head, eyes would dilate, whiskers
would flair and then rotate forward, and the hair on her tail and back
would stand up like a Halloween cat. Even without a hiss or growl
(she didn't do either) there was no doubt when trouble was coming.

Ginger had attacked her owner a number of times on his bed,
where the cat slept at night, once even leaping at his face and biting
him on the lip. I instructed him not to let the cat on the bed at all until
she became less aggressive. "Put your feet under the covers and just
encourage her to get off," I told him. Mr. Lewis somewhat reluctantly agreed to this precaution. I was surprised he hadn't decided on
his own to keep Ginger off the bed when she started lunging at his

face, but I have long since ceased being amazed at what owners are willing to put up with from the pets we love.

We did the typical displacing, directing the aggression away from the victim to something else such as a toy. We talked about how the owner could reward Ginger with treats for docile behavior during the day, and I recommended that he ask the veterinarian for a more appropriate medication for Ginger when they had to travel— one with less sedative but more antianxiety effects, which has shown good results for traveling cats.

I strongly urged Mr. Lewis to invest in a cat carrier, a simple step that might have prevented a great deal of aggravation had he done it before the first trip. Like children without seat belts, small animals loose in cars are a risky business. Cats usually love being in containers—nearly every cat owner has found his pet curled up inside a box, drawer, or grocery bag—and will quickly adapt to being confined during a car trip, particularly if the cat is properly introduced to the carrier before the trip.

The treatment program in both these trauma cases hinged on the use of cat cages or carriers. I know that getting some cats into carriers can be a traumatic event in itself. The fact that the carrier is associated with the dreaded *visit to the vet* makes it all the worse from the feline's point of view.

I've had clients whose cats went berserk when the carrier was taken out of the closet, and clients who needed first aid after being clawed and bitten by their terrified pets as the cats were pushed into the carrier. That's the only way the owners had ever been able to get them in.

If you and your cat cringe at the thought of the cat carrier, you might do some desensitizing exercises before the next trip to the vet or elsewhere.

The first step is to bring the carrier out of the closet and casually leave it around the house for a few days in various locations. Your cat might be a little alarmed at first, because chances are, he normally sees the container only when he's about to be shoved into it and then taken out and given a shot.

Let him sniff it and so forth, and then put a couple of cat treats inside to lure him in. After a couple of days the cat should be jumping in and out of the box to get his treats. (As your cat makes progress, tell him how great he is. Praise helps.) I know of previously box-phobic cats who by the third day of this procedure have no problem lying patiently in the box waiting for the owner to toss in some goodies.

When you reach this point (it may take less time or longer than three days) the next step is to pick up the container and carry it for a couple of feet, then a few yards, then across the room—depending on the cat's comfort level. Then repeat the routine with the lid closed. Don't move too quickly from one stage to the next. If he jumps out of the carrier en route, just let him go and try it again later. The idea is to make the whole experience as pleasurable and nonthreatening as possible.

Eventually you will be able to close the lid of the case and actually take the cat out of the house, because now being confined predicts pleasant things to him. Just make sure to give him plenty of praise and healthful treats.

In the case of the terror-on-wheels cat, Ginger, she became so attached to her carrier that she began sleeping in it in the bedroom at night, which helped to end the possibility of any aggression when her owner was in bed. And she eventually was able to ride in the carrier in the car, like any other cat.

You can't always prevent traumatic events, although you can check out your house for cat booby traps to minimize the risk. And you can't predict how any one cat will react to life's little (and big) accidents. But if the accident happens, and the cat doesn't seem to be back to normal within a few days or a week, you'll probably need to desensitize the cat to whatever has scared it. Take it slow and easy, and have patience. It may take some time, but there's a good chance your cat will get back to normal with your care and understanding—and professional help, if necessary.

12

Hatred, Revenge, and Inflexible Felines

" I was so humiliated I could have died," cried the tearful voice over the phone. "This is positively *ruining* my relationship with my fiancé. Why did Figaro *do* that to me?" She paused for breath.

"This kind of thing does happen now and then," I answered. "It must have been quite a shocking experience."

"It was *awful!*" she wailed, picking up steam. "We had just about opened our eyes in the morning, and Figaro hops up on the bed, looks Ted right in the eye, turns her back and just *lets loose*, all over his chest! I can't understand why she would hate him so. Now he doesn't even want to come over here."

I sympathized with the young woman and told her I would call on her the next day. As I took down the address, it seemed to be a familiar one.

"And you just moved there from out of state?" I asked.

"Yes, just six weeks ago, and I have a new job, and then I met Ted at work, and we've been making plans, but now Figaro is going all over the place, besides spraying my boyfr—"

"Does your apartment have a Pullman kitchen and a pink marble bathroom and a garden out back full of yellow rose bushes?"

For the first time she was speechless. But she recovered quickly.

"Well, yes. Have you been here before?"

"Yes, I have. About five months ago I had a client at your address."

Now I could picture the damp, musty little basement apartment. A cat named Tucker had absolutely refused to use the litter box that his owner wedged under a shelf in the pantry. Whenever my client would bring the cat there to remind him of the box, Tucker ran away. All the feline's deposits were made smack in the middle of the kitchen floor, where he could see (and get away if startled) in all four directions.

It had been one of my simplest cases ever.

"Your cat is trying to tell you that he needs a better view of things and some escape routes when he goes to the litter box," I had told the owner. "Right now he feels trapped in the pantry where he can't see a thing."

"Oh."

"You don't have to put the litter box in the middle of the kitchen floor, but let's pick a spot that has a little bit more of the feeling Tucker is looking for."

We moved the box to the top of the stairs, and he never failed to use it again.

"The last cat that lived in your apartment had a litter-box problem, too," I told my new client, Mindy Allen, "and as I recall, the apartment is somewhat musty."

"It stinks," she sniffed.

"Maybe some of those leftover odors are contributing to the problem. We'll go over everything together tomorrow."

When I met with Mindy the next day, she had calmed down and was ready to work on the cat's problems. First, she brought me up to date on all the changes that had taken place for her pet and herself in the past few weeks. Besides moving across the country, the cat had had to adjust to a new apartment and litter box, a new schedule in which her owner—a computer saleswoman—now had very irregular hours, and an abundance of odors left by our friend Tucker.

On top of everything else, Figaro—a pretty white longhair sporting a rhinestone collar—was playing unwilling hostess to a nasty collection of fleas, which are always bound to add to a cat's behavior problems. And the last straw, apparently, was the addition of Ted with his own special scents. The cat did not seem to be high strung or particularly stressed out (she responded with only a curious look when I clapped my hands), but she was definitely unhappy. She wandered around crying frequently during my visit. Perhaps the fleas and the move were influencing both the spraying and the house-soiling problem.

As we discussed the cat's toileting habits, an odd fact was revealed. It seemed that the cat would use the litter box (in the bathroom) only when her owner was also answering nature's call. Whether this meant that the event had social overtones for the cat, or that it was the one time she felt sure that she would not be startled in the act, I don't know.

To cover both possibilities, I asked Mindy to move the box away from the door to the other side of the toilet, where the cat would feel less liable to be surprised.

Then I instructed Mindy to buy another box to put in a separate location in the apartment, so that Figaro would have one totally unconnected with Mindy's bathroom activities. This box would be for Figaro to use while Mindy was at work.

Mindy had to get rid of her pet's fleas and institute a new policy of "no cat in the bedroom," which was a big plus in luring the disgusted fiancé back to the apartment.

I explained to Mindy that Figaro was not trying to do something mean to her, and didn't "hate" the new boyfriend. She was just having difficulty organizing her activities around her new location, Mindy's comings and goings, and all the different new odors.

"Some cats have a hard time coping with changes, and yours has had quite a few of them," I told her. With defleaing, the masking of odors in accident areas, and the help of a temporary antianxiety drug, Figaro was soon using both litter boxes and having only an occasional accident. After six weeks the follow-up phone calls stopped.

Unfortunately, less than four months later, Mindy and Figaro (now with Ted) were on the move again, transferred by the employer to a new location.

How did I know? Because that was when I got a *third* call from that same address. A new tenant in the apartment contacted me after Mindy had broken her lease and moved away. I was beginning to wonder if my number was written on a wall somewhere in the building.

"My vet recommended I call you," she began. "It seems—"

"Don't tell me," I interrupted her. "You have a cat that's not using the litter box."

"That's right," she confirmed. "I didn't realize Dr. K. had spoken to you."

He hadn't, of course.

It had to be something about that apartment.

Mindy's lament that Figaro hated Ted and that her misbehavior was something the cat was "doing to" her is a familiar theme when I deal with inflexible felines. I had another case in which the cat would pounce on the new boyfriend's neck and bite him; that client, too, was convinced that the cat despised the interloper. In these cases I hear a lot of the following:

"I know he's doing it to get back at me!"

"She's punishing me because I . . ."

"He always sulks when we . . ."

"She didn't want to move away from Pittsburgh, and that's why she's . . ."

All perfectly logical sounding "explanations" for irritating, repetitive cat misbehavior. What these cats have in common is that they don't adapt well to change, but the owners often believe that the misbehavior is directed at them. Sometimes people feel ambivalent or guilty about making the changes, asking themselves, "Are we doing the right thing?"—and seize upon the cat's action as "proof" that they're not.

Alan and Clara Langer were no exception when it came to telling me what was wrong with their cat, Goldie, a five-year-old purebred male. They knew exactly why he was acting like that.

"Alan's new job requires a lot of travel," Clara explained as she served me coffee in her high-tech, all-white kitchen. "More than I'm happy with," she added confidentially, "but Al and I can live with it."

Alan walked in and shook hands. "Goldie's the problem," he began. "He hates it when I leave town, so he sprays in the house, and when I come back he tries to punish me by spraying my suitcase."

"How does Goldie do while Alan is gone?" I asked Clara.

"He's fine after a day or so. It's only when Alan comes back that he's upset. It's like, 'if you're going away, don't come back!'"

"I'm sure it must seem like that's what Goldie's trying to tell you," I said with a smile. "How does Goldie behave the rest of the time, when Alan isn't traveling—any litter-box problems?"

"Oh no," Alan replied, "he's a good cat ordinarily. Always uses his box. In fact he still uses it on the days he sprays. Right, Clara?" His wife nodded.

"The vet gave him a tranquilizer to calm him down when Al leaves," Clara added, "but it seemed to make him so drowsy we stopped using it."

"That's a common reaction for the first couple of days of the medication," I told her, "but if the dosage is correct, cats usually adjust to it."

They had probably stopped before Goldie got used to the drug. At any rate, I decided to recommend a low dose of a female hormone to calm Goldie during the crucial periods. I told Alan and Clara that I would go over the medication with the vet to make sure it was the proper medicine and dosage for Goldie.

"Now, what do you do if you catch the cat spraying or find evidence that he has sprayed?"

"We sometimes rub his nose in it," said Alan.

"But if I see him getting ready to spray, I can sometimes stop him by clapping my hands or tossing him a toy," his wife added.

"OK. I'm going to ask you to stop any physical punishment you may be trying. Talk to him as calmly as you can and try not to shake him up." I had observed that though he seemed quiet, Goldie was easily startled. Usually, if cats are fairly calm and confident, you can slam doors and drop dishes and they won't react noticeably. But this one jumped when the kitchen timer buzzed.

"Now, would it be feasible to block off his access to some of the house during these departure and arrival periods?" I felt that the cat would be less liable to be aroused if he had a smaller space in which to organize his activities; the less stimulation, the less arousal and spraying.

"Sure," said Clara, "I don't need to use the dining room or even the living room while Alan is away. I can put up some baby gates to keep Goldie out of there. He's not much of a jumper," she added, gazing down at the portly pet lying at her feet.

"OK," I said. "Now, I want you to understand that your cat is probably not trying to punish you for traveling, Alan, or telling you not to come back, or anything like that."

Goldie's owners exchanged sheepish glances.

"I know it's hard not to take his behavior as a message when he goes all over your suitcase, but his spraying is just a logical reaction to something novel—your suitcase—coming into the house with many strange smells attached to it."

I told the couple that they seemed to have a very inflexible cat that just couldn't cope well with changes in his daily life. When Alan took a trip, Goldie was unable to remain calm, and then by the time Goldie got back to normal, Alan came home and upset the cat again.

"So what you interpret as Goldie saying, 'Now I'll show you!' is actually, 'I was able to cope with this change once, but when you come back, it's more change, and I can't cope.'"

"So we shouldn't take it so personally?" Clara asked with a grin.

"That's right."

"Well, be that as it may," Alan said, "we still can't have this cat wrecking our house. If he doesn't stop, we're going to have to do something drastic."

"We couldn't give him to anyone else if he's so sensitive to change, could we?" mused Clara. "He'd likely have the same problem with them."

"Or worse," I guessed. "I understand how upsetting this problem is, but I believe it can be handled pretty well by the vet prescribing a maintenance dose of medication every other day, increased slightly before departure and arrival days by dosing twice daily on those days. I don't like to suggest using this long term, because there can be some serious side effects."

"They can't be as serious as having him put to sleep," Alan remarked.

"Well, let's see what we can do for Goldie," I said. "Please stop rubbing his nose in the urine—it isn't helpful at all. But distracting him with toys or food or playing when it looks as if he wants to spray is a good idea. Keep doing that."

So we began a program of no physical punishment, medication adjusted to periods of change, distraction before spraying, a restricted environment, and—not least—a clearer understanding of Goldie's point of view.

After a few weeks of the program, Goldie virtually stopped

spraying. Mindful of long-term effects of the medication, they tried decreasing the dosage after about two months of successful use, but the cat immediately started spraying again when Alan left town. So he went back to the minimum dosage that would handle the problem. Considering the alternative, it was a risk the couple and the veterinarian were willing to take.

Why should a cat be so inflexible? Cats organize their behavior around certain daily activities. All kittens do this, but some are more adaptable than others to changes. And some cats get so locked into a pattern of activity that they become almost compulsive. If they are thwarted in any way, they get upset.

And getting upset leads to arousal, which leads to spraying, or urinating or defecating outside the box, or other misbehavior. And that's when the owners call me. Typically these problems will show up fairly soon after something has changed; they don't take long. As Goldie did, the cat might wait until you are back home from a trip before getting into trouble. And that's when people interpret the inability to cope as revenge. "Well, you're finally back from Denver—you'll be sorry now!"

What activities and patterns are so important to a cat? A young cat might like to lie in the dining room window to catch the morning sun. Next, she gets a cat treat from her owner after he finishes the laundry. Then she waits for the mail to come flying through the mail slot. After that she goes to the bedroom window for the afternoon sun and watches for the kitten that comes to explore around the back porch. Later, the cat may go to the top of the steps when she hears the school bus coming to greet the kids. And so forth.

While the routine might vary slightly, the cat's behavior becomes organized around these stimulus events that become almost synchronized with her activities; thus we can predict what the cat is going to do day after day, week after week, month after month, year after year. (Keep a "cat activities" journal for a week if you want to tune in to your cat's worldview.) All evidence suggests that the cat does not get bored and want to remodel the kitchen, move to a new house, or do anything else drastically different—provided these events and what she does in reaction to them become organized early in the cat's life.

As you can imagine, household moves present a huge challenge to the inflexible feline. I handled a case in which a cat owner had moved to a larger unit within a condominium complex just two buildings away. The owner was concerned and frustrated because his indoor/outdoor cat had returned to the doorstep of the former unit every day of the three weeks since the move. When he looked for her, she'd be patiently sitting there waiting to be let into the old apartment. I had to summon all my reserves of tact to explain to the owner that the cat wasn't "stupid" or "dense," as he characterized her, but was apparently more attached to the old location or territory and daily activity pattern than to the hand that fed her.

This client was lucky the old apartment wasn't far away. If the owner of a cat who is reluctant to move with the household doesn't force the cat to stay inside for a while and reorganize his daily pattern in the new place, the cat may well continue to go back to the old location. This is not characteristic of all cats, but there are more cases of it in cats than there are in dogs—enough to retell "Homeward Bound" many times over.

Sometimes when people's pets are made to go along for the ride, they can end up as battered as a well-traveled suitcase. What may be a series of great adventures to a carefree single or an energetic family can turn the ordinary house cat into a basket case, especially if the cat was not particularly adaptable to begin with.

The unfortunate cat that comes to mind was a four-year-old domestic long-haired male named Ali. By the time I met him, he had logged more miles on the road than most salespersons.

His kittenhood began quietly enough. At a very early age he

found a happy home with Dottie Ellis, with whom he lived on the East Coast for two years. Then Joe Gilcrest came into their lives. And Ali started moving all over the nation with both of them.

I don't know what Mr. Gilcrest did for a living—that's something I don't ask a client—but whatever it was, it made putting down roots an impossibility. First the trio moved to the Pacific Northwest. I wanted to know how Ali reacted to this crucial initial move.

"Not too well," Ms. Ellis reported as we met at their current residence. "Basically he hid under the bed for a year."

"I see," I said. Not an auspicious first reaction to change.

"Then we moved to the Midwest and stayed in three different hotel apartments, just for a month or so in each, and the maid service there sort of drove him up the wall. He never did get used to those different women coming in every day."

"Then what?"

"Then we finally moved here, about three months ago, and Ali's been causing a lot of problems and we're really, like, bummed out. We even got him a kitten to play with, but Ali was so hostile we finally took it back to the shelter."

I looked up at my client. It was natural for her to blame her pet for his inability to adjust, I supposed, but it was hard for me not to feel sorry for the cat, who was hiding under the bed as we spoke.

"Ali's had to adapt to a lot of changes in his life," I remarked. "What exactly are the problems he's having now?"

"Well, first off, he hates this wonderful house! Can you imagine?" Her gaze swept around the brand-new colonial. "He won't step a foot in certain rooms. I can't even get him into the kitchen to eat!"

"What other areas does he avoid?"

"Mostly the bathrooms, the front hall, and the guest bedroom."

I asked her to show me the rooms. While most of the house was covered in mauve wall-to-wall carpeting, the spare bedroom had hardwood floors. And the entranceway, kitchen, and bathrooms were finished in marble and ceramic tile.

"I'd like you to think back to the other homes and the suites Ali has lived in," I asked her. "Did they all have carpeting, or has he encountered smooth, hard surfaces before?"

"Well, I don't know." She thought a moment. "I've never had tile—much less marble. We've had linoleum or indoor-outdoor in the kitchens and always wall-to-wall carpeting or area rugs in the other rooms."

"Do you think Ali is unhappy about the feel of those floors? He's used to the soft carpeting."

"Aha!" the young woman exclaimed, grinning. I was glad to see her gaining a better understanding of why the cat was not behaving the way she and her spouse wanted it to.

Clients sometimes need to know that they can make a difference, that neither they nor the cat is crazy and that they have the answer to a problem all along—all I did was enable them to see it. Just as in the management of a business, progress in treatment goes well if the owners are directly involved in the process.

"This is very typical of inflexible cats," I went on. "If the surfaces don't feel like the kinds of materials they're used to, they just refuse to have anything to do with those areas—even if it means not eating."

"I catch your drift. But this tile was imported from Italy, and I'm not going to cover it up with anything!

"I understand. We'll work around that—in fact we can use Ali's particular nature to our advantage. You say he's having litter-box problems?" She had mentioned spraying as well, and I had noticed a number of cats outdoors as I swung into her driveway.

"He's having accidents in a bunch of places, including in the potted plants." She rolled her eyes. "I read somewhere to put mothballs in there, but it didn't work."

"Let's look at those spots and see how we can make them more novel and unattractive to him."

Ms. Ellis showed me around the house, and I talked to her

about putting plastic carpet runners over the spots the cat was using for toileting. The carpet runners had the kind of hard surface Ali had avoiding walking on in the house.

Then we talked about some ways to make the litter-box area more attractive so that he would want to use it regularly, and selected an alternate spot for the cat's food and water dishes.

"The neighborhood cats Ali can see outside your house are probably contributing to his spraying," I remarked. "Can you possibly speak to the owners about keeping them away from here?"

"Oh, sure," she replied. "All those cats belong to Lily next door. She's cool; she already asked me if her pets were a nuisance in my yard, and I said no because I thought they were keeping Ali company."

"What they're keeping him is overly stimulated," I remarked. "If she's willing and able to get them away from here it would be a good idea."

Then we turned to the next problem.

"The other thing is that Ali follows me around all the time," Ms. Ellis complained. "Except when my husband is home, and then he hides under the bed. Once or twice Ali tried to bite him."

"Does he bite you?"

"Never."

I queried her further and found that the cat and Mr. Gilcrest had very little to do with one another.

"Do you think your husband would be willing to take on some of the responsibilities of feeding Ali, changing his litter, and playing with him?" I asked.

"Well, I don't know," she replied frankly. "Joe always refers to Ali as 'your cat.' "

"I think that's part of the problem," I answered. "I'll bet Ali would be more comfortable if he was able to include dealing with your husband in his bunch of preferred activities. He's probably not too happy spending his time under the bed."

"I'll ask Joe."

Ask him she did, and Mr. Gilcrest began to take a more active role in caring for Ali. The next-door cats were kept under better control by the neighbor, who now let them out only briefly in the morning. As my clients settled in to the house, accepting Ali's preference for carpeting and helping him with the litter-box problem, the cat began to gradually adjust to his new situation.

A month later, his owner reported some definite improvement, and announced that they were going to stay put this time, at least for "a while."

"That's great," I replied. For Ali's sake, I hoped so; he had already made enough moves to last the rest of his nine lives.

If you are facing a move, try to plan for your cat's adjustment to the new situation. You might be wise to use the same steps I outlined in Chapter 1 for introducing a kitten to a new home. If your cat seems to enjoy keeping track of everything that goes on in your (his) home, allow him to view the packing and loading and unloading and unpacking processes. These cats seem to thrive on participating in the move.

But if you have any concerns at all about your cat's sensitivity to change, it would be best not to ask him to relocate with the family during that crazy, stressful time called "moving day." Try to board him at a kennel or with a friend for a day or two, until everything is moved into the house and basically put in place. This way, his first introduction to his new home can be a calm and pleasant one, and if he has a negative reaction to the few days of boarding, he at least won't relate the unpleasant experience to his new address. If you pay attention to your cat's personality, you'll be able to choose confidently between boarding him and keeping him with you during the move.

Again, moves and other changes are not going to upset every cat; some felines are so temperamentally calm they seem to be able to cope with just about anything. Few are like Mitts, the most stubborn, inflexible cat I've ever come across—too inflexible to cope with my fairly routine treatment program for inappropriate elimination.

Mitts was a big male Himalayan who had taken to defecating on the bunk beds in a bedroom of the large house in which he lived with his owners and their three kids. The two boys were threatening not to sleep in their beds anymore by the time the owners called me. The last straw was when the cat defecated on the owner's bedspread as well—though the litter box was only a dozen feet away in the adjoining dressing room.

When I went to the house, I found a fairly typical situation. The litter box was in a very poor location as far as privacy and escape potential were concerned, so I recommended that the client move it to a better place in the room and temporarily add a litter box in the boys' bedroom, where Mitts was showing a preference for doing his toileting.

I also advised the owner to put some cat toys on the boys' bunk beds and have them play together there, in order to transform them in Mitts's perception into play areas. We made a few more minor changes, and I felt confident that there soon would be some definite improvement in the situation.

The case seemed straightforward. But I made one big mistake. I hadn't asked Mitts for his opinion of the treatment program. He soon let on what he thought of it.

Four days later the client called to inform me that the cat had not defecated on the beds. That was the good news. The trouble was, he had not gone in his box, either, nor anywhere else. Since my visit, the cat had spent a fair amount of time lying next to the litter box, staring at it, but would not use it.

I recommended that Mitts see the veterinarian, who administered an enema, to everyone's relief. But five days later, the cat had still not moved its bowels on its own. And, the client told me pointedly, in her opinion Mitts was worse off now than before she had consulted me. I had to agree. I spoke to the vet, who had seen the cat again and X-rayed it for any intestinal blockages. There was no apparent physical reason for the cat's retention of its stool. The vet and I agreed that this was one inflexible feline, so sensitive to change that he would rather just not go than go in a box that had been moved a distance of fifteen feet.

I had no choice but to put the box back where it had been and use another method to make the beds unattractive to the cat. Sure enough, Mitts promptly went back to using his box, and fortunately, the accidents in other locations became fewer. Mitts resumed sleeping on my client's bed, virtually ignoring the boys' room. In fact, though we kept a box there, I was told he never even entered the boys' room again. After a couple weeks, the client grudgingly admitted that the cat was now (finally) better off than before she had consulted me. Things were looking up.

Then I received another call. The misbehavior had resumed, "for no reason." Nothing had changed.

Nothing, I finally extracted, except a bad snowstorm that had caused the household to lose electricity for two days. That's when the family moved downstairs to sleep around the fireplace. The storm didn't seem to bother Mitts, except that he had refused to budge from the foot of the mother's bed those nights, even though he was sleeping all by himself in the now-frigid room. And the accidents started again.

But surely this couldn't have anything to do with his naughty behavior, pronounced the caller.

"We know Mitts doesn't handle change very well," I reminded my client in a vast understatement. "I have a feeling that the disruption in his routine caused by the storm may have made him regress temporarily."

"Just because of the blackout?" she asked skeptically.

I explained again how a disruption of a stable pattern of behavior introduces stress, which leads to dysfunctional behavior. And one of the ways that we return the cat to good behavior is to reintroduce stability and a routine of doing the same things in the same way, day in and day out. It would take some time to get Mitts back to his routine, but I urged the owner to be patient with him after this stressful period.

"By the way," I asked, "how did the rest of you cope with the lights and heat being off?"

"It was awful," she answered promptly. "Very stressful."

Maybe cats aren't so different from people after all.

13

Babies, Kittens, Dogs, and Other Surprises

Babies, divorce, new kittens, dogs. Big changes such as these can upset even the most flexible cat. Let's start with the baby.

Sometimes in the excitement of preparing for a new addition to the family, the resident cat is all but forgotten. A family that did things right when bringing a kitten home from the shelter may not give a thought a few years later to preparing the same cat for a new baby (or kitten or puppy).

We've seen that cats can be extraordinarily inflexible. They tend to be suspicious of anything different or novel, and for the most part, don't care for change. New babies, kittens, puppies in *my* house? No, thank you! Or everybody fighting all the time, and then I have to move? No way! Felines are susceptible to the same stresses that humans are and need us to smooth the way for them as much as possible. Here are a couple of suggestions that can help the cat accept the inevitable.

Before bringing a newborn home from the hospital, give the cat access to the nursery and to small blankets or articles of clothing with the baby's scent on them. On the big day, allow the cat to sniff the baby in your presence. It's best to supervise access of all pets to the baby for the first few weeks, allowing them to get used to the baby while someone familiar and trusted is in the room.

No, cats do not suck the breath from babies (a surprising number of people still believe this myth), but a cat may be curious enough to jump into the cradle or crib, frighten the baby, or accidentally scratch it; or just bring in dirt and bacteria on his feet, or a flea in her coat. A cat also may be attracted to a baby because of the nice milky smell (less so with formula) or because lying next to one provides her with the warmest place in the house. Keep in mind, too, that some cats knead with determination, and a baby's body is fair game. Being allowed supervised access to the baby will help the cat accept and become acquainted with the small intruder with a minimum of distress.

Try to give the cat as much attention as possible before and after the baby's birth. Kitty will often take a back seat to the new arrival, and if this means the reduction or elimination of play routines, it will be one more unpleasant reality for the feline to adjust to. Add the lack of attention from the family to the new alarming sounds emanating from baby, and things may start to go haywire, as one set of new parents I called on recently noted with alarm.

"I'm worried, Dr. Wright," confided Eleanor Reston, the young mother who had greeted me at her door with an infant wailing in her arms. "In the three months since Belinda arrived, Stripe has been acting really strange—he either ignores the baby or acts as if he hates her. It's reached the point where I'm afraid to leave the two of them alone together." She shook her brown ponytail in dismay. "Why hasn't Stripe gotten used to our little girl?"

As she quieted the infant with a pacifier I reassured Mrs. Reston that I was confident we could help Stripe adjust to Belinda. "First, I'd like to understand some of the changes your cat's been reacting to.

So let's go back a few months and recall for me what life was like before Belinda came along, and how it's different now."

As Mrs. Reston's story unfolded, it proved to be all too typical. Understandably, the couple's firstborn child had stolen the spotlight from their pet, and the new dynamics of the family had caused changes in the way the cat behaved. Stripe, the pampered pet and center of attention for seven years, was feeling the stress of having routines and relationships turned upside down. Now the declawed tiger-striped male was focusing an increasingly malevolent gaze on the source of all his misery—little Belinda (whom the huge cat still outweighed by a couple of pounds).

The cat seemed to "look right through" anyone who was holding, feeding, diapering or doing anything else for Belinda, the new mother told me, and would "glare" at her infant daughter in a bold and aggressive way that she and her husband found unnerving.

"Eleanor's not exaggerating," said Mark Reston as he came into the room and shook hands with me. "There've been times when we thought Stripe was about to bite or pounce on Belinda. I know this must sound weird," he added uncomfortably.

"Not at all," I assured him. "Your pet seems to be viewing the baby as an unexpected source of competition for important resources—your attention and his space. You know, cats will often stare at other cats, as you've described, to make them go away."

"Great," he replied gloomily.

"Since Stripe and your daughter are about the same size," I went on, "Stripe may be dealing with Belinda more as though she were another animal of some sort than as a small human. He may know Belinda isn't a cat, but he doesn't know any other way to deal with an intruder that size. This is mildly threatening behavior, and it's better to stop it before it escalates. So you were right to be concerned."

I waited a moment for them to digest this information, then asked them a question.

"Have you ever seen Stripe go into a room where Belinda is on the floor or in her baby seat and plop himself down very close to her but with his *back* to her?" I asked the couple.

"*Yes!*" they answered in unison. Mrs. Reston smiled a little as she gently placed the infant in her husband's arms.

"He does that all the time—when he's not staring at Belinda, he absolutely refuses to face her. . . . That's what I meant when I told you that Stripe ignores the baby," she told me.

"Oh, okay." I grinned. "Shall we say, *pointedly* ignores?" They nodded as I made an entry in my notebook. "I've seen several cases where the resident cat treats a newcomer this way," I told them. "I've even seen tiny cats do it to huge dogs in homes where the cat is the boss, apparently with no fear of getting stepped on or attacked. Although we can't know for sure, it may be Stripe's way of express-ing his superiority over Belinda."

"Wait till she starts pulling his tail," Mr. Reston said darkly.

"Any other behavior related to the baby?"

The duo told me that the cat would not go into the nursery, and that when the baby came home with them from the hospital, Stripe had abruptly stopped sleeping between them in their bed and now slept in the hall between the nursery and their room. When they car-ried the cat to their bed Stripe sometimes stayed there but didn't seek out his former favorite sleeping place.

"And Stripe used to come when I called him," added Mrs. Reston. "He still might come, but now he's just as likely to hiss or growl at me."

Stripe also was not using the litter box consistently and was hav-ing accidents in the house. This had never happened before.

On top of all this, Stripe became very fearful and hid or ran to the back door when the baby cried, which was pretty frequently.

"OK," I said, making a note of these behaviors. "By the way, where is Stripe right now?"

The couple looked at each other and then gazed vaguely around the room, which was littered with toys and baby paraphernalia. "He's probably behind the dryer or standing at the back door because the baby was crying," Mr. Reston finally guessed.

"Belinda must keep you pretty busy," I commented to the two of them as they fussed over their little one. "You probably don't have much time to play with Stripe anymore."

"None," retorted Mrs. Reston bluntly. "I don't even have time to let Stripe in and out now, much less play with him."

"So the litter box is new?" I inquired.

"Oh yes," she replied, "it's new for the daytime. Stripe used to go to the bathroom outside. He preferred it. I would watch him from the window or go out with him, so that he could come right back in after a few minutes.

"We don't like him to stay outside long because of the animals in the woods around our house. But I don't have time for that now. Excuse me—" She hurried off to diaper the baby.

The solutions to the problems in this case were relatively simple, perhaps facilitated by the drastic change the cat had undergone in his relationships with his people. Stripe was sure to notice the difference when the treatment program began.

Both Mr. and Mrs. Reston were probably totally ignoring the cat now that they had a baby. Unfortunately, it's par for the course in a lot of families, and when problems develop, it points to the importance of social interaction for cats with family members.

My first "prescription" was for them to play with, praise, and handle Stripe whenever he was around the baby, at least several times a day—something they used to do routinely when there were just the three of them. That way Stripe could associate the baby with positive things, not with negative ones like being ignored or being in competition with the infant. Soon the cat would predict that good things would happen to him in the baby's presence, and he would begin to actually prefer the baby being around.

For the fear caused by Belinda's crying, I taught Mr. and Mrs.

Reston how to start a counterconditioning program similar to the ones we use for dogs that have phobic reactions to thunderstorms. They were to tape-record the baby crying and invite Stripe onto the couch where they would stroke and pet him as they played the tape at a very low volume. This was to be done four or five times a day for short periods. As they were petting him, the cat would associate the low-volume baby cries with being stroked and petted. Gradually and over a period of time, the volume would be turned up, until it was the same volume as the baby crying, and the pet was still relaxed.

This procedure is called systematic desensitization. The animal can't run away and be calm at the same time, so this is the way we enable the cat to experience calmness, in the presence of the same stimulus (crying) that used to result in running away.

Other animal behaviorists have found that clients prefer to teach their cats to avoid the baby's room, to avoid potential interaction with the infant. If parents want the baby's room off limits to the cat, creating an aversive or unpleasant experience when the cat tries to enter can prompt the cat to ignore the room and not even attempt to enter it.

In this procedure, a novel item like a closed umbrella or new plant is placed at the entrance to the room. At the same time, a loudspeaker or other sound device is set up to startle the cat when it approaches. (Pet stores and catalogs now carry products that emit a loud tone, activated by motion inside the doorway.) The cat sees the umbrella and associates it with the loud, unpleasant noise as it nears or enters the room.

After one or two trials, the cat should avoid entering the room when the strange object (the umbrella) is present, even without the noise. It's the same principle as leaving the waxed paper on the kitchen counter after the mousetraps are removed.

I don't advocate this procedure, because it doesn't seem like a nice thing to do to a cat, but some cat owners have apparently found it to be very effective. A potential negative side effect, though, is that the cat may come to hide under the couch most of the day and never get used to the baby.

Stripe's litter-box problem was solved by moving the box from upstairs into the attached garage and putting a cat door in between the kitchen and the garage. This provided enough of an outdoor feeling to motivate the cat to go out to use his box. It also gave Stripe a private place to escape to when all the fussing over the baby became too much for him.

Although they were very busy with the baby, my clients took the few minutes a day the program called for and put it into action. Within one week, they were on the phone happily telling me about Stripe's progress. Stripe was now actually seeking out Belinda and facing her without staring or glaring. He was back in his old favorite bed, and he didn't race for the door when the baby started to cry, but more or less ignored it. Stripe was happy again, they said. They sounded happier, too.

"We just didn't realize what a big effect Belinda's birth would have on Stripe and how we were neglecting him," Mrs. Reston said in her final feedback phone call.

"Well, don't be too hard on yourselves," I told her. "I guess we need to start up some sibling classes for pets when a new baby is about to come home, just as hospitals do for big brothers and sisters. Children are rarely unprepared for a baby's arrival anymore, but there are no classes to help cats and dogs adjust."

She laughed in agreement.

And, although we don't talk about jealousy, perhaps our household pets are more "human" than we animal behaviorists like to admit—or are able yet to understand.

What about the other side of the coin—when the family is breaking up? In addition to the stresses of moving or having household routines change when it comes time for parting of the ways, the tensions generated by the divorcing parties often filter down to affect the pets. Cats and dogs have even become key ingredients in custody battles. In one recent case a Hollywood star sued his live-in lover for custody of the two cats he had brought to the relationship and she had taken away with her. The court deposition regarding the cats ran to 750 pages.

Unfortunately, cats don't always come out unscathed and ready to be divvied up like coffee grinders and living room sofas when family relationships are crumbling. This was never clearer than when I was called to an impressive designer's showplace to deal with a problem of "cats fighting." But that turned out to be only a symptom of some real trouble in paradise.

The people and animals I found inside this custom-built home apparently hadn't been living happily together for a long time. Whether the owners were driving the pets nuts, or the pets were driving the owners nuts, was never exactly clear to me. All I know is that when I rang the doorbell, I stumbled into the middle of a situation involving three very expensive cats, two of which were fighting and spraying all over the house, pedigrees be damned; one large, unruly, tail-thumping dog; and two pissed-off (and occasionally -on) human beings who had already filed for divorce.

The tension in the house was palpable, and all the inhabitants—including one ghostly white Persian I barely caught a glimpse of—were reacting badly to everything. I sat down with the husband, Mitchell Brown (his wife wouldn't stay in the same room), and tried to start sorting things out.

The focus of Mitchell's ire seemed to be Sasha, a ten-year-old female Persian who ruled the roost. At least she spent a great deal of time and energy trying to. Most people tend to think that only male cats spray, but, as Sasha's discouraged owner related, here was one spayed female who sprayed up a storm.

She also attacked Star, a four-year-old Persian Mitchell characterized as "overly sensitive," and the only male pet in the house. The third cat—the white one—spent most of the time slinking around trying to avoid getting caught in the action.

The dog, a Collie named Tiffany, seemed oblivious to all the feline intrigue, but the atmosphere in the house probably contributed to her rather hyper behavior. She raced around administering enthusiastic licks to anything within range, which frequently proved to be a little cat's face or rear end. This added to the pets' irritation with each other and with life in general.

"And Sasha hates me," Mitchell added glumly as he wrapped up his list of complaints.

"What makes you say that?" I asked.

"Well, she only sprays on *my* side of the dresser in the bedroom, and only on *my* mirror."

"And what do you do when this happens?"

"Well, I usually throw something at her."

"And if you don't catch her in the act?"

"Usually my wife Tracy will find the wet spot later. It could actually be on a glass or marble surface anywhere in the house—except on *her* stuff—and Tracy'll go get the cat and I'll smack her in the face."

When someone tells me something like this, I try to probe and see if he is just using an expression or is actually striking the cat, and in what manner. I've found that "smack it in the face" can mean anything from shaking a finger at the cat to tapping it on the nose to actually slapping its face. Unfortunately, in this case, Mitchell readily admitted to hitting the cat with his hand.

With that kind of response, it wasn't too difficult to understand why Sasha "hated" her owner, and why she continued to spray after the heightened stress of being disciplined. Yet Sasha followed Tracy around constantly, Mitchell told me, and tried to prevent her rival Star from getting close to the woman.

I explained to Mitchell that the cat would not understand why she was being hurt since there was no way she could connect this punishment with what she was being punished for, and that he and his wife must stop all physical retaliation immediately. Then I went to take a look at the horrible troublemaker.

Sasha was one macho female. She strutted around, very stiff-legged and full of bravado, yet if someone sneezed or dropped a fork or laughed suddenly (though there wasn't much laughing going on in this house), she would just about go through the roof. The cat would leap up, shoot across the room, look wildly around, and then gradually try to resume her "queen of the realm" pose as if nothing had happened to startle her. She might then begin to lick herself as if that was all that mattered in the world (a "displacement" activity, no doubt, to give the impression that "I'm in complete control"). From my perspective, Sasha was a pathetically anxious feline.

It was pretty clear that what with trying to keep control of three other pets and two humans (through spraying or following) and having to deal with cat fights, flying shoes, and smacks in the face, the villainous Sasha was carrying a lot of anxiety around with her as she patrolled her turf.

The false front Sasha put up showed me she was using all the coping mechanisms she had. Her behavior was so purposefully assertive that the next little thing that happened would likely exceed the cat's ability to cope. I'd seen this phenomenon before—cats who are severely stressed out but act laid back, as if they haven't a care in the world. Although these cats will go to great lengths to appear as though nothing is wrong, they are often right at the threshold of losing it, and with the next thing that goes wrong, they do. Because cats with timid personalities have a hard time exploring and dealing with novelty, they are especially likely to have difficulty handling stress.

I have to empathize with cats like Sasha. Some days I make the mistake of seeing too many difficult clients, and then I walk into a home like this one, where everybody is yelling at everybody else, and it seems an impossible task to start setting things back on the right track. That's when I'll find myself sitting in the corner of a couch, looking very relaxed and talking calmly to the client, and I'll notice that my hand is clamped on the arm of the couch with a grip that turns my knuckles white.

So at that point I just try to relax and take a deep breath, and go on. But I don't think it's so different with cats sometimes—they try to look cool, but you know that they have "white knuckles" too.

At any rate, for his part, Star was fighting back when stalked by Sasha and spent the rest of the time lying on surfaces he or she had sprayed—a habit foreign to most fastidious felines unless things are quite out of hand. In this case Star apparently needed the sprayed-on spots as "lookout points" when trying to keep track of his attacker. It seemed as though my work was cut out for me.

I wanted to analyze the locations where the cats spent the greatest amount of time fighting, and Mitchell immediately pointed to the bedroom. The biggest trouble spot in the house was the bed, which he had eventually vacated as the marriage fell apart.

Now that empty spot right next to Tracy had become the most prized, sought-out, and fought-over spot in the house for Sasha and Star. It was the scene of frequent attacks; when he went to shave in the morning, Mitchell would often find Star quivering in his sink after losing a nocturnal bed-space battle with Sasha.

The bed problem was the most obvious one to lay directly at the feet of divorce. Everyone recognizes the potential impact of marital distress on children, yet the effects of family dysfunction on household pets have been swept under the rug, along with the broken china.

I wasn't sure how much I could help the pets, as long as the tension in the household remained unresolved. But the fact that the owners were willing to work with the animals was a good sign. I would give it my best shot.

First, I suggested putting a chair with a soft cushion next to the bed for one of the cats to sleep on, making Tracy the middle person at night, with the cats separated but allowed equal access to her as she slumbered. With more choices, the cats would now be able to settle on which one slept where, ideally with a minimum of fur flying. We selected a chair that was lower than the bed, to help maintain whatever semblance of a dominance relationship the two cats had, with the top cat claiming access to the bed.

The next battle site was in the pantry. The cats had to jump up on the counter to eat, all three of them fighting over one big bowl of gourmet cat food and water. I pointed out to a chagrined Mitchell that the local alley cats probably found less competitive conditions out back at the garbage cans.

We then turned to consider the litter box, which was situated six or seven feet from the food dishes. Again, there was just one box for three cats, and it was near the dog's food and water dishes.

In this ultrasmart, spacious home, the four animals were forced to compete for everything they needed for existence—all in the space of a small back hallway. They had the best food money could buy and their water and litter were sufficiently clean; it was just that the arrangement was a recipe for disaster for this particular quartet of troubled pets. When I pointed this out to Mitchell, he agreed to make a few simple changes that could help lessen the opportunities for conflict.

First, the situation called for more than one litter box, preferably in an area of greater privacy.

"Would you be willing to install a kitty pass-through in the door that leads from the kitchen to the basement?" I asked. "This will give Star and the white cat a chance to get away from Sasha and avoid ambushes at the litter box—not to mention giving the cats a place to escape from your dog." He nodded.

Similarly, two more feeding stations were added so that competition over food would not continue to add to the cats' troubles.

The odor of the urine spray on many of the marble and glass surfaces in the house would keep the cats aroused, so I told Mitchell to use some mild creamy chlorine cleaner on those spots. Finally, I suggested a much-needed temporary tranquilizer for both Sasha and Star, to which their owner readily agreed.

These improvements were able to bring some measure of relief to this beleaguered household. But it was not until the divorce was declared final, and Sasha and Star were split up along with the sterling silver, that these pets caught in the middle were able to really relax. I believe Mitchell ended up with Star, the dog, and the house. Tracy took the white cat and Sasha. The cats were able to lead more or less normal lives again, although Mitchell reported that Star never gave up sleeping in the bathroom sink—just in case.

Sasha, Star, and company offer a good example of the typical headaches cat lovers in homes with more than one cat often face. Owners whose motto is "the more the merrier" even after reading Chapter 7 should proceed with caution.

The introduction of a kitten to a resident cat, often done with the admirable motive of providing the cat a friend or playmate, can be just as distressing to the resident cat as adjusting to a human baby—or more so. Throughout this book you'll find plenty of examples of clients I've seen because their cats can't get along. So aggression toward the new cat, or sometimes from the new cat to the resident cat, is a standard reaction. But another, far less prevalent reaction to be aware of may be one of distress or sadness that is initiated or magnified by the introduction of a kitten.

Especially where the rationale for getting the second cat is to cheer up the first one, I'd be very careful, because this well-meaning strategy often backfires. Such was the case with Bandit, a two-year-old Maine Coon Cat, who had seemingly lost his zest for life and his appetite. When the vet could find no physical cause, Donna Todd decided to bring a female Maine Coon Cat kitten into Bandit's life to "cheer him up."

This well-meaning gesture set the stage for disaster. Bandit quickly became more lethargic, his appetite worsened, and then both cats became ill. Still no physical malady could be identified. Although the kitten recovered quickly, poor Bandit had to be given fluids intravenously and was left weaker than ever.

Now, weeks later, Bandit was still weak but had received a clean bill of health from the vet. Although he was back on his feet, the male had a new problem. His new pal, Ember, had begun threatening or actually attacking him when he approached his food bowl; occasional play biting had turned into frequent ambushing around the feeding station. And once that happened, his owner told me, Bandit simply wouldn't go near the bowl the rest of the day.

Although Maine Coon Cats are among the largest and heaviest breeds around—weighing up to a hefty thirty pounds—Bandit tipped the scales at a puny twelve pounds when Donna called me in. She seemed to have a legitimate cause for concern.

"You're sure Bandit is healthy now?" I asked. "His physical condition isn't likely to be giving him trouble in some way and causing him to avoid his food dish?"

"No, we've been back and forth to the vet several times," she replied. "He's supposedly all well now."

Bandit lay limply in a corner of the dining room as we talked, while Ember amused herself with a toy mouse.

"I'm sorry Bandit is having such a rough time," I told my client.

The young woman nodded miserably.

"What steps have you taken to try to deal with this situation?"

"Well, mainly I've just tried to keep them separated most of the time," said Donna. "I have Ember's box and food upstairs in the bedroom, and she stays around there, and Bandit has his food and litter box down here. But I'd rather they were friends."

"That would be nice," I agreed. "But let's make sure we help Bandit with his problems first. He's not a happy camper."

I told Donna that my diagnosis was behavioral depression.

"Bandit's depressed?" she asked. "Why?"

"That's the big question," I replied. "Sometimes dealing with a lot of changes can trigger something like this—just as it can for people."

Donna looked troubled. "Bandit has had some changes to cope with in the last year or so," she said. "There was the divorce, then we moved here, and when he stopped playing I thought a kitten would help. Another big change. I should have realized it was a dumb id—"

"There's no way of knowing what combination of things have upset Bandit," I interrupted before she could lay a heavy guilt trip on herself. "The cause of this isn't really so important—I can see that you love your cat and you've done the best you could to cope with everything."

"I guess so," she replied listlessly. The young woman seemed a bit low herself.

"At any rate, let's see what we can do to get Bandit up to speed," I said. "And with any luck, he and Ember will begin to have a better relationship. Maybe she'll pick on him less when he stops being such a good target and gets back to his old feisty self."

I wrote down the following instructions:

1. Work on stabilizing Bandit's daily routine. Do things in the same order, at the same time each day.
2. Reward Bandit for social activity with Ember. Praise him and play with him when Ember is nearby. Gradually increase the amount of time Bandit and Ember spend together.
3. Don't force Bandit to do anything he doesn't want to do.
4. Allow Ember to get up on something (stool, nightstand) to look out the window in her room. This will divert her attention from Bandit and allow the cat to keep track of things outside, which may help reduce the general stress level.
5. Keep track of whether Bandit resumes eating his favorite foods as time goes on.

6. Confirm that the vet checked for any problems causing discomfort (gums, teeth, etc.) to rule out physical cause for Bandit's lack of interest in eating.

I also told Donna that I would like to recommend a tranquilizing drug for Bandit to take temporarily while he relearned appropriate behavior.

Why a tranquilizer?" she asked in surprise. "He barely moves now."

"I think the proper dose of the right drug will relax him and enable him to start being better able to cope with Ember," I explained. "And there's an extra benefit to a side effect from the drug I have in mind."

"What's that?" Donna asked.

"Given the proper dosage, it increases the appetite."

"Great!"

The tranquilizer enabled Bandit to regain his interest in food and the much-needed extra energy gave him the boost he needed to solicit play from Ember. As their relationship stabilized, we were able gradually to wean the older cat off the medication.

And so with a little help, Bandit came to act like a regular cat again, and at the same time grew accustomed to his new "friend."

If it's a dog you are welcoming into your household containing felines, you're probably going to have an easier time of it. Unless the dog is notoriously aggressive to cats, the two will probably get along fine—once the cat shows the dog who's boss.

The old saw "fighting like cats and dogs" gives a faulty impression about the ability of our favorite household companions to peacefully coexist. The two species are in no way natural enemies, and as millions of pet lovers can attest, many cats and dogs are the best of friends. (Probably one-third of all cat households have a pet dog as well.)

Sure, an unrestrained dog might chase any cat who's willing to run across the backyard. But what happens in the unlikely event that the dog catches up with the cat? A defensive swipe of the claws in the vicinity of the dog's nose usually puts the issue to rest at once. The same thing is likely to happen in the home if the dog is overzealous in his sniffing or licking, or if—to the cat's way of thinking—the dog is generally making a nuisance of himself. (I'm not talking about dogs with aggression problems. In those cases, you'll need to protect the cat and consult an animal behaviorist to help the dog.)

After one or two such confrontations (even if the cat is striking out with a declawed paw) man's best friend usually gets the message and settles into a respectfully subordinate role. This makes for some strange scenes in a cat-and-dog home. I've seen big panting canines sitting waiting their turn with dry tongues hanging out longingly while the household cat laps daintily at the dog's gigantic bowl of water. And similarly, if a treat is tossed to the cat, the dog waits with great restraint until the cat walks away before going over to lick up any remaining crumbs.

Of course, there are exceptions to this usual state of affairs. One of my clients could see disaster looming from such a distance that she called me in for some preventive counseling on how to introduce her new dog to two equally new cats. And with good reason. I could understand someone adopting a Greyhound dog just retired from the track. These hard-working animals surely deserve to spend their remaining years with a loving family.

What was rather curious was that the same family would then turn around and adopt two little gray furry animals that bore a striking resemblance to the objects the dog had been bred and rigorously trained to chase. But the deed was done and there was no turning back.

The Windham household already contained a sweet elderly

Labrador in addition to the new Greyhound when the teenage daughter brought home the two eight-week-old tabbies. The Lab had outlived two household cats that had "raised him" as a puppy, but the greyhound was not likely to be so welcoming to the kittens. No one had the heart to make the daughter return the two balls of fluff she had fallen in love with, but Travis and Erin Windham dreaded what would happen when the kittens met the racing dog. So Daffy and Tulip were holed up in April's bedroom until I could get there and figure out how to keep Trump from eating them alive.

The kittens had been born into a household with children. So they had been handled frequently and were great with people. But they had never seen a dog before. So when Ranger, the Lab, came up to say hello and lick their little faces, the kittens loved it. That must be what dogs are like. Bring on the Greyhound!

Mrs. Windham wasn't willing to do that. Controlling Ranger during his introductions to the kittens was one thing; she was afraid the trusting kittens might not have the chance to escape if Trump went for them. But the kittens couldn't stay in the bedroom forever. That's when the family called me for help.

Trump wasn't stupid. He knew the kittens were upstairs; for one thing, he could smell them. Mrs. Windham explained that they had taken the dog upstairs, just to see what he would do, and put a muzzle on him—just in case—even though the daughter's door was shut. "But that seemed to get him very excited and upset," she reported, "so we brought him back down."

Well, if you've ever been to the dog track, you know that when Greyhounds race, they run with muzzles on. So this was like a "context effect" for the dog. When the muzzle was put on, rather than functioning like an inhibitor as it would with most dogs, it actually *prepared* the dog for chasing. So they had inadvertently sensitized Trump to get into the chase mode. With the odor of the kittens in his nostrils, he must have been ready to come out of retirement then and there.

I had my clients put the muzzle away, and we got down to business. Right away the dog showed me some good signs. The retired Greyhound had some favorite toys he would carry from room to room with a very gentle mouth—things like shoes, tennis balls, and stuffed animals. He would collect and put these toys on the Oriental rug in the sunroom of the large Victorian house. The dog was allowed to rest on the couch above his favorite objects.

But though he was gentle with his toys, Trump was fairly aggressive to some dogs; that was one of the things that concerned us—he might likewise show aggression to the little furry kittens. We couldn't predict whether he would see the kittens as toys he could be gentle with or more like the "rabbits" he'd spent the better part of his life chasing around the track.

The last thing I wanted to do was to expose the dog to the kittens abruptly, and I reassured the Windhams that they had done the right thing in keeping them apart thus far. My plan was to use a variety of senses to desensitize the dog to the cats very gradually, before he actually approached them. By the time the dog met the cats, he would be used to the type of sounds, smells, looks, and movements he was about to encounter.

We started out by letting Trump go up to the bedroom without the muzzle on and without the kittens there. This way he could sniff around and say, "Aha, here's something that's a familiar smell." But it wouldn't be associated with anything running around. We'd let him get habituated to the odor of the kittens for the first week.

In the sunroom, I wanted him to lie down in his usual relaxed position as his owners put a few things the kittens used, including a soft blanket, on the couch next to him so he could sniff them while he was calm. This was a way of systematically desensitizing him to the kittens' smell so that it would automatically be associated with being calm and relaxed.

If Trump could be relaxed in the presence of the kittens' odor on the blanket, then that was one step closer to being relaxed around the actual kittens. This was the first step and one that was most likely to succeed. I usually try measures that are most likely to succeed to start with and then move on to more and more risky types of exercises.

At the same time, we wanted to provide Trump with something he could chew, such as a rawhide bone, so that if the dog did get some kind of arousal from the scent of the kittens, he could redirect it onto the bone, rather than looking for a kitten to chew on. Fortunately, he was very relaxed during this period of time, and he desensitized well to the odor of the kittens.

Next we tried to desensitize him to the appearance of the kittens. Here we ran into some good luck. We knew from doing a good behavioral history—which is always a necessity in these cases—that Trump had shown no interest in chasing squirrels, even though squirrels dart around and bear some resemblance to rabbits. When the Windhams had taken the dog to the wooded backyard, where the squirrels hopped all over the place, Trump had just looked at them briefly or ignored them. And we knew that the "rabbits" he chased around the track were white, whereas these kittens were gray.

I had Mrs. Windham go to a toy store and get a stuffed gray kitten. If Trump would associate or generalize the toy kitten to the squirrels, then he wouldn't be interested in chasing or biting the stuffed kitten either. So in the next week, we brought the toy kitten to the couch and had him look at and sniff it, and later on we rubbed the real kittens' towel, loaded with their scent, on the stuffed animal.

We also put the stuffed toy in the kittens' room overnight. They played with it and slept on it, and then we returned it to the Greyhound the next day. Now the dog was desensitized to something that looked and smelled like a kitten and was gray like the squirrel and the kittens. And the dog was still calm.

So far, so good. We then added a little movement, because the kittens tumbling about might tempt him to attack. So as we relaxed him on the floor in the sunroom and stroked him, we tossed the toy kitten from one person on the couch to the other while saying "good dog" in a soothing voice, trying to keep him in a relaxed mood.

Then we had April tape the kittens vocalizing for about a minute, and exposed Trump to the tape. Since the miaows didn't presage anything very interesting—just more tossing back and forth—the dog quickly lost interest in the kittens' sounds. Purrfect!

After about ten days, it was finally time to introduce the kittens. We had April hold them in her lap while the dog relaxed on the floor with a leash on. (No muzzle.) Then Trump was allowed to walk around the couch, and while they had a leash on him, the dog was allowed to go up and meet the kittens.

Trump sniffed the kittens, intensely interested in finding out what they were all about. If it looked as if he was starting to get excited, Mr. Windham pulled the dog back with a "down, stay," and then a quiet "good dog." This made it seem as though behaving calmly around the kittens was Trump's idea, and he was consistently rewarded for it. After only three or four repetitions, Trump began to merely sniff the kittens or ignore them altogether.

The next step was to let the kittens wander about on the floor. This worked out fine. The very gradual procedure of systematically exposing the dog—and the dog's senses—to things that were more and more like the real kittens had succeeded in suppressing any interest in chasing them. When the kittens became unthrilled with having this big dog in their faces, they started to avoid Trump, and jumped up on things to distance themselves from him. And the dog never did pay much attention to the cats (perhaps they were too much like squirrels). So they coexisted without incident, just the way most ordinary dogs and cats do the world over.

We moved very slowly and carefully with this pair of new kittens, but the same principles apply to introducing new kittens or cats to a resident pet in any household. If you follow these steps and gradually introduce the pets to each other in a controlled setting (controlling the dog, allowing the kitten to reciprocate an approach or escape), everything should work out fine. Just don't toss them together cold; that's asking for trouble and can be frightening or unsettling for both old and new pets.

The Greyhound introduction had a happy ending, as do most cat-meets-dog stories. The majority of such introductions don't require any intervention at all, beyond what I've described. There are, in fact, no serious dog/cat behavior problems, with the exception of coprophagia—the habit many dogs develop of cleaning out the cat's litter box. (Dog owners know that their pets may also seek to ingest dog, deer, or other animal feces.) Solution: Keep the litter box where the dog can't get to it—in an elevated spot or in a room accessible only by a cat door. You don't need a cat therapist for that one.

It doesn't take much to make a cat and her owner happy together. Good food, a soft bed, toys, some turf of her own, some stability, a sunny windowsill, and a caring human being are the simple requirements for a long-term, satisfying relationship. And what these animals give back to us is worth more than anything money can buy. So take good care of your precious cats and if you ever think they're acting crazy, don't be afraid to ask for help. They're worth it!

Suggested Reading

Bradshaw, John W. S. 1992. *The Behaviour of the Domestic Cat*. U.K.: C.A.B. International.

Hart, B. L., and L. A. Hart. 1985. *Canine and Feline Behavioral Therapy*. Philadelphia: Lea & Febiger.

Leyhausen, P. 1979. *Cat Behavior*. New York: Garland STPM Press.

Marder, A. M., and V. L. Voith, eds. 1991. *The Veterinary Clinics of North America, Small Animal Practice: Advances in Companion Animal Behavior*. Vol. 21. Philadelphia: W. B. Saunders Co.

Polsky, R. H. 1991. *User's Guide to the Scientific Literature on Dog and Cat Behavior*. Los Angeles: Animal Behavior Counseling Services, Inc.

Tuber, D. S., D. Hothersall, and V. L. Voith. 1974. Animal Clinical Psychology: A Modest Proposal. *American Psychologist* 29: 762–766.

Turner, D. C., and P. Bateson, eds. 1988. *The Domestic Cat: The Biology of Its Behavior*. Cambridge: Cambridge University Press.

Voith, V. L., and P. L. Borchelt, eds. 1982. *The Veterinary Clinics of North America, Small Animal Practice: Symposium on Animal Behavior*. Vol. 21. Philadelphia: W. B. Saunders Co.

« »

To obtain a list of applied animal behaviorists certified by the Animal Behavior Society, write to John C. Wright, Ph.D., Chair, Board of Professional Certification, Psychology Department, Mercer University, Macon, GA 31207.

Index

Index

Birds, 86, 113
Biting, 10, 16, 27–28, 85, 105, 126, 148, 154, 161, 176, 191
  ankle, 14–17, 61–63, 101, 102, 126, 128, 148
  hand, 115–17
  infections from, 105–106
  punishment for, 61–63
  tail, 14–16
  treatment for, 106–18
Boarding, 192
Brain, 17, 79, 113
Breed, 8, 24–25, 118
Breeding cats, 22–24
Burmese cats, 77

Carpet, 101, 189–90
  scratching, 91–95
  urination on, 10, 34, 41, 46, 49, 51, 53, 55, 60, 70–72, 73–75, 87–88, 161, 165–68
  vomiting on, 70
Carriers, 174, 177–78
Car travel, 31–32, 175–78
Catsitters, 35, 51
Centers for Disease Control and Prevention, 105
Chaining behavior, 146–47
Chemosignals, 103–104
Children, 2, 8, 11, 35, 105, 148, 169, 196–202
  newborn, and cats, 196–202
Choosing a cat, 23–40
Claws, 91–95, 159, 176, 177
  declawing, 92, 93, 135, 168–69
  scratching problems, 91–95, 168–69
Clay litter, 48–50
Cleanliness, litter-box, 42, 51–53, 57, 154–55, 159
Clicker, 96, 97, 98
Colitis, 165–66
College cats, 31–35

Conditioning behavior, 83–99
Constipation, 164
Crying, 142

Declawing, 92, 93, 135, 159, 168–69
Defecation, 93
  on bed, 193–94
  inappropriate, 31–35, 41–59, 61–65, 66, 67, 87–89, 106, 119, 129, 148–55, 158–69, 180–94, 199, 202–208
  pain-related, 164–69
Defensive postures, 107–108, 117–18
Depression, 209–11
Diabetes, 164
Diarrhea, 164, 166
Digestive disorders, 161–66
Diseases, 69, 164–69
Divorce, 196, 202–208
Dogs, 1, 2, 11, 15, 20, 54, 75–76, 83–84, 92, 102, 105, 107, 120, 128, 135, 196
  bites, 105
  relationship with cats, 39, 203, 204, 207, 211–17
Dominance and subordinance, hierarchies of, 128–35, 203–208
Door, cat, 86, 140, 202
Door banging, 137–39, 162

Ear mites, 161
Ears, 27, 85, 107, 135
Eating habits. *See* Food and eating habits
Elimination, 36, 39, 40, 41–59
  and health problems, 164–69
  inappropriate, 2, 6, 9–10, 19, 31–35, 41–59, 61–65, 66, 67, 70–72, 73–75, 87–89, 106, 119–25, 129–33, 141–42, 145–47, 150–53, 154–55, 158–69, 180–94, 199, 202–208

Index

Index